TEACH YOURSELF
THE BASICS OF
ASPEN PLUS™

TEACH YOURSELF THE BASICS OF ASPEN PLUS™

RALPH SCHEFFLAN
Chemical Engineering and Materials Science Department
Stevens Institute of Technology

AIChE®

WILEY

A JOHN WILEY & SONS, INC., PUBLICATION

A joint publication of the American Institute of Chemical Engineers and John Wiley & Sons, Inc.

Published by John Wiley & Sons, Inc., Hoboken, New Jersey.
Published simultaneously in Canada.

For general information on our other products and services or for technical support, please contact our Customer Care Department within the United States at 877-762-2974, outside the United States at 317-572-3993 or fax 317-572-4002.

Wiley also publishes its books in a variety of electronic formats. Some content that appears in print may not be available in electronic formats. For more information about Wiley products, visit our web site at www.wiley.com.

Library of Congress Cataloging-in-Publication Data:

Schefflan, Ralph.
  Teach yourself the basics of Aspen Plus / Ralph Schefflan.
     p. cm.
  Includes bibliographical references and index.
  ISBN 978-0-470-56795-1 (pbk.)
  1. Chemical-process–Computer simulation. 2. Chemical process control–
    Computer programs. 3. Aspen Plus. I. Title.
  TP155.7.S28 2010
  660′.280113–dc22

                                                            2010019514

Printed in Singapore.

oBook ISBN: 978-0-470-91006-1
ePDF ISBN: 978-0-470-91004-7
ePub ISBN: 978-0-470-92285-9

10 9 8 7 6 5 4 3 2 1

To Ruth

CONTENTS

viii CONTENTS

14   COMPLEX EQUILIBRIUM STAGE SEPARATIONS   199

INDEX   213

PREFACE

During my years working as a chemical engineer in development laboratories, process engineering groups, and plant startup and support operations, the most frequently referenced documents were process flow diagrams (PFDs), which contain the material and energy balances and the basic process design information. Equally important were process and instrument diagrams (P&IDs), which contain details of all equipment, all controls, all instruments, and all lines (i.e., process, instrument, and utilities). Process simulation software is an excellent tool for producing high-quality PFDs, and when integrated with computer-aided design software, facilitates the production of P&IDs. There are several process simulation software systems available to the chemical engineering community, and Aspen Plus is arguably the most popular.

Teach Yourself the Basics of Aspen Plus™ evolved from two graduate courses that I taught at Stevens Institute of Technology over the past 20-odd years. The first course, ChE662, is an introduction to steady-state chemical process simulation, which is usually taken by graduate students and is organized around a series of workshops that introduce Aspen Plus functionality. Occasionally, undergraduates are enrolled and typically experience difficulties in the thermodynamics of phase equilibrium and parameter estimation, due to limitations in their undergraduate courses. The second course, ChE612, deals with the analysis and design of complex equilibrium stage processes and with difficult multicolumn problems such as, extractive distillation systems. Over time, the course evolved from the use of stand-alone two- and three-phase flashes, decantation, and two-phase distillation software, to their equivalent blocks in Flowtran and later, Aspen Plus.

The idea for this book originated from my observations of students in these courses. I noted that after an initial period dedicated to learning the basics of how to navigate, locate material, and enter data into Aspen Plus, students could proceed through the exercises, within the workshops, mostly on their own. I would give an introductory lecture for each subject studied, show examples, and provide individual help on the workshops when needed. It was a rare student who did not finish all of the workshops during the course. The book is organized in the same manner,

If you expect to teach yourself Aspen Plus by reading this book, you will be disappointed. Aspen Plus is a complex process simulator, and in my opinion, the best way to learn is with hands-on experience, by attempting each exercise within each workshop, and when difficulties are encountered, referring to the problem setup and solutions on the accompanying CD as well as the notes in the workshop section at the end of each chapter.

The accompanying CD contains the input and solutions to all the examples and workshops. There is a root folder for each chapter, within which there are subfolders named Examples and Workshops. Each example and workshop exercise is provided in .bkp, Aspen Plus format and .txt format. The .bkp files are set up as input files to view details, and may be executed. The .txt files are solutions and may be viewed with Notepad by a reader who does not have access to Aspen Plus. References to material on the CD within each chapter of the book are by subfolder/filename: for example, Chapter Four Examples/Rubin. Some of the workshops were developed using earlier versions of Aspen Plus, and when attempting to use them from the CD, a message to that effect may appear; however, all have been executed successfully with version 7.0, which currently resides on the server at Stevens Institute of Technology. I recommend that while reading the text, Aspen Plus be used simultaneously to execute and review each example. If Aspen Plus is not available, the .txt solutions may be reviewed.

The book is designed to be used by undergraduates, graduate students, and practicing chemical engineers. The first section of the book explains the basic structure of the software and leads the student through a hands-on introduction to the various features of the software designed to facilitate the setup of simple problems. Features such as the material-balance-only option, access to Aspen Plus documentation through Help, the Next button, menu navigation, and the report function are introduced. The remainder of the book is organized in a series of sections that focus on particular types of operations: for example, a two-phase flash. Each chapter is accompanied by the equivalent of lecture material that describes the equations being solved, various limitations, potential sources of error, and a set of workshops containing exercises that the students should solve to gain experience with the particular subject. Some of the exercises are designed to produce errors that students need to analyze in order to complete their experience. Much of this part of the book is suitable for undergraduates, although some will be limited by courses in their curriculum that have not yet been taken (e.g., exposure to the thermodynamics of phase equilibrium). Undergraduates should limit their exposure to Chapters Six and Fourteen. Chapter Six deals with phase equilibrium and provides exposure to the most popular thermodynamic equations as well as material on parameter fitting. Chapter Fourteen addresses advanced problems in distillation. Graduate students and practicing engineers who undertake these sections should have had exposure to undergraduate equilibrium stage operations and, preferably, a graduate course in thermodynamics.

This book is not intended to be a self-study guide for all the features of Aspen Plus. For example, material on some reactor blocks, batch blocks, and the thermodynamics of electrolytes is not covered. Many subjects not addressed can be found by selecting the Help button on the main Aspen Plus display. The philosophy of the book is based on the idea that once a chemical engineer becomes thoroughly facile in the use of the software and has a good understanding of the basic blocks, he or she should be able to learn to use many of the unaddressed functions by applying the same philosophy as that of the text itself: namely, to study appropriate sections in chemical engineering textbooks that describe the subject matter and to familiarize oneself with the function's implementation by reading Aspen Plus's documentation and attempting a sample problem. As an example, to understand the Aspen Plus electrolyte methodology, it would be useful to read the section on electrolyte equilibrium in *Molecular Thermodynamics of Fluid-Phase Equilibria*, 3rd ed., by J. M. Prausnitz, R. N. Lichtenthaler, and G. M. de Azevedo (Prentice Hall, 1999) and in Aspen Plus's Help, and follow up with the section entitled "Generating Electrolyte Components."

I have made an effort to provide the describing equations of most of the models (blocks) referred to here, and if not possible because of the proprietary nature of the software, I have described the functionality. One should recognize that Aspen Plus is proprietary software and that the source code and implementation details are not available. Additionally, there are frequently several ways to solve the equations that describe the blocks, and there is no way to ascertain these details since Aspen Technology does not provide them.

I wish to acknowledge the help provided by Aspen Technology's academic support group, especially for the loan of an Aspen Engineering stand-alone license for use while I was out of the United States and unable to access the Stevens Institute of Technology server.

RALPH SCHEFFLAN

CHAPTER ONE

INTRODUCTION TO ASPEN PLUS

Aspen Plus is based on techniques for solving flowsheets that were employed by chemical engineers many years ago. Computer programs were just beginning to be used, were of the stand-alone variety, and were typically used for designing single units. The solution of even the simplest flowsheet without recycle required an engineer to design each unit one at a time and, manually, introduce the solution values of a previously designed unit into the input of the next unit in the flowsheet. When it became necessary to deal with a recycle, the calculations began with a guess of the recycle values, and calculations ended when the values produced by the last unit in the loop matched the guesses. This involved much repetitive work and convergence was not guaranteed. This procedure evolved through the construction of rating models of units, as opposed to design models, which could be tied together by software in a way that emulated the procedure above and employed robust mathematical methods to converge the recycle elements of the process. This type of system is termed a *sequential modular simulator*. An excellent example of such software was Monsanto Corporation's Flowtran (1974), which eventually became the kernel upon which Aspen Plus was built.

Subsequently, Aspen Plus, although still basically a sequential modular simulator, has grown considerably and has many advanced functionalities, such as links to a variety of specialized software, such as detailed heat exchanger design, dynamic simulation, batch process modeling, and many additional functions. It also has a facility for using an equation-based approach in some of its models, which permits convenient use of design specifications in process modeling.

The Aspen Engineering Suite, which incorporates Aspen Plus, can be installed in a variety of ways using network servers or on a stand-alone personal computer.

Teach Yourself the Basics of Aspen Plus™   By Ralph Schefflan
Copyright © 2011 John Wiley & Sons, Inc.

Installation is the responsibility of either the user, with tools provided by Aspen Technology, or the information technology department, which services the user. This is done only once and modified, typically annually, with future releases. Whether the user's installation is by network downloads or by CDs, it is necessary that the user select desired modules, at a minimum Aspen Plus and its required add-ons and associated documentation. No other modules are necessary.

1.1 STARTING ASPEN PLUS

When the Aspen Plus User Interface icon on the desktop is clicked or, alternatively, the sequence All Programs/Aspen Tech/Process Modeling V7.0/Aspen Plus/Aspen Plus User Interface is selected from the Start menu, the Aspen Plus Startup display shown in Figure 1.1 appears.

A selection from a list of existing applications may also be made, or other stored files may by selected by choosing the appropriate radio button. If the Template button is selected, a list of application-oriented possibilities is shown (Figure 1.2). Although the flowsheet selection appears in the Run Type box in the lower right-hand corner of the display, other selections are available from the associated drop-down list.

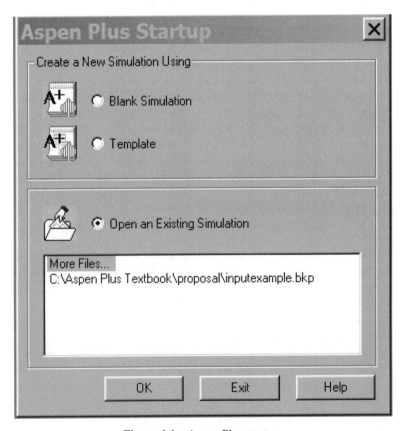

Figure 1.1   Aspen Plus startup.

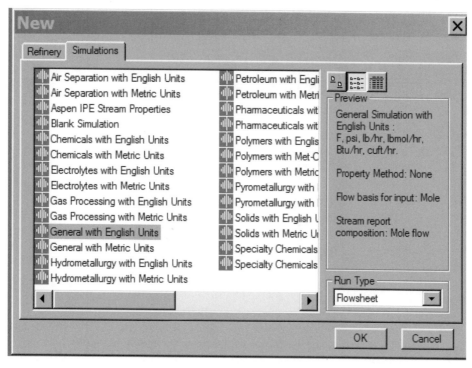

Figure 1.2   Preconfigured selections.

If the option "blank simulation" is chosen, an application can be custom configured. After selecting the desired option and clicking OK, the Connect to Engine screen appears, and upon selecting OK, a blank workplace that facilitates the graphic users interface appears if "flowsheet" was chosen. If the option chosen is anything else, the first required input form appears.

1.2   GRAPHIC USERS INTERFACE

The graphic users interface (GUI) is the means by which a flowsheet is defined. The process consists of placing blocks and streams on the workplace and connecting them. Aspen Plus assigns generic names such as B1 to the blocks. The user may change these names by right-clicking on the element of interest and using the menu that appears. Blocks are selected by choosing a category tab from the model library—for example, Mixers/Splitters—and clicking on the icon that represents the block desired. After clicking, a movable + sign appears on the open area of the display. After positioning it on the screen, a left click will place the block. The + sign remains and can be moved to insert another instance of the same block. This function ceases when the arrow button at the lower left is selected. In identical fashion, streams can be placed on the flowsheet. Material, heat, and work streams may be selected. When a stream input is selected and the cursor is moved onto the workplace, the ports to which the streams may be connected are shown. The connection is made by moving the active

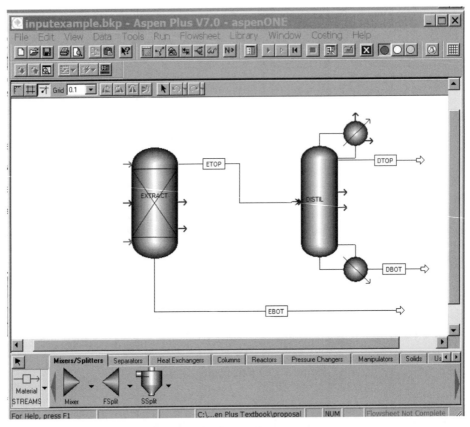

Figure 1.3   Connecting streams.

cursor over an open port and clicking. An example of connecting streams to ports is shown in Figure 1.3.

All icons, block names, and stream names can be selected and moved using standard Windows techniques. Similarly, streams can be moved, rerouted, disconnected, and reconnected. Selecting and right-clicking on any of the objects displays a menu that provides many useful functions for manipulating the graphics. These include changing icons, rotating objects, renaming, deleting, and aligning the graphics.

1.3 NEXT BUTTON

Aspen Plus provides the user with a mechanism for filling out forms in an orderly fashion. At any point after the flowsheet has been fully defined with the GUI, the user may select the Next button, which appears as the symbol N $\rightarrow$ on all forms. The Next button moves to the next form required. On occasion, after using the Next button, Aspen Plus will prompt the user to select from a choice of actions to be taken. The Next button provides only the minimum required input. As an example, when an activity coefficient equation to be used in the simulation is chosen, Aspen Plus will use a default data source, such as a vapor–liquid equilibrium (VLE) source; however, if the simulation involves liquid–liquid equilibrium (LLE), it is the users' responsibility to select the appropriate data source. Aspen Plus will not open the appropriate displays by using the Next button.

1.4   SETUP SPECIFICATIONS DISPLAY

After the flowsheet has been defined with the GUI, pushing the Next button brings up the Setup Specifications—Data Browser display. The Data Browser panel of the display should show a list of all possible categories that can be chosen for selecting various options and for entering appropriate data. If the setup information shown in the left panel of the display does not appear, choosing the upper menu selection, Data, and entry Setup will present the Data Browser panel, shown in Figure 1.4. If the list of process model icons shown at the bottom of Figure 1.4 does not appear, clicking the View menu Model Library will display them.

When starting a problem, the Setup Specification display provides a drop-down list associated with the entry box Run type, which shows the six primary functions that Aspen Plus is capable of:

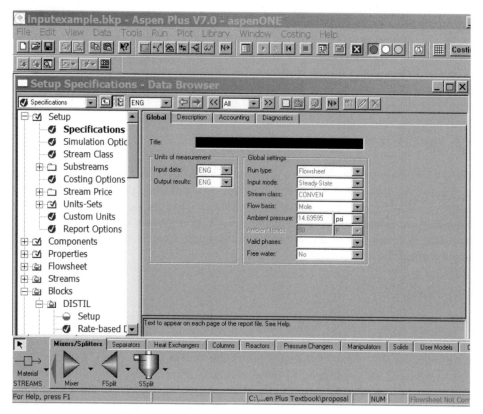

Figure 1.4   Setup specifications for data browser.

1. *Data regression:* fitting data to models

2. *Flowsheet:* process simulation

3. *Property display:* showing properties of a component in Aspen Plus's database

4. *Property analysis:* estimating physical and thermodynamic properties

5. *Assay data analysis*

6. *Property plus*

The last two functions are not considered here. The user selects the required function to initiate data input for a specific requirement.

Note that the Data Browser panel shows a blue check next to various items. This indicates that the default values are sufficient to proceed with data input; however, this is the minimum data required and the values may be modified to meet the requirements for any problem. The red elements on the browser panel list indicate that user input is required. A red element may be deleted by right-clicking on it and selecting Delete from the menu that appears, or if Aspen Plus permits, and sometimes by using the Windows delete key.

There are drop-down lists for all the categories in the data-entry boxes under the Global tab, and all may be changed. For example, the value Mole has been selected for the option Flow basis. Additionally, under units of measurement, the input and output units have engineering (ENG) as default values, while the associated drop-down list offers Metric and SI units as other possibilities.

Aspen Plus provides a Help button, on the topmost menu, for accessing information by subject. Additionally, help with any entry on any display is available by moving the cursor to the entry and pushing F1 or selecting the ↖? button on the tool bar.

1.5 SIMULATION OPTIONS

If the Simulation Options category is selected, the Simulation Options display shown in Figure 1.5 appears. All of the default values that appear under the various tabs need not be changed except under the tab Calculations, where the option not to use energy balances in the calculations is available. This is an important option for preliminary calculations. If simulations do not involve solids or electrolytes, the appropriate options may be unchecked.

1.6 UNITS

Aspen Plus provides a user with a choice of units: engineering, metric, and the international system of units, SI. An important option is the ability to select mixed units; for example, the choice of engineering units with mmHg and degrees C as temperature and pressure is not uncommon in some pharmaceutical applications. To accomplish this, Units-Sets is selected from the Data Browser menu, which produces the Object Manager shown in Figure 1.6. Then, selecting New produces an Identifier for the new unit set, in this case US-1, and a choice as to whether or not to assign this as a global data set. Then Figure 1.7 appears.

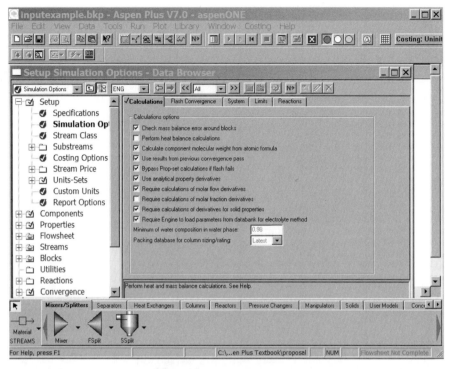

Figure 1.5   Simulation options.

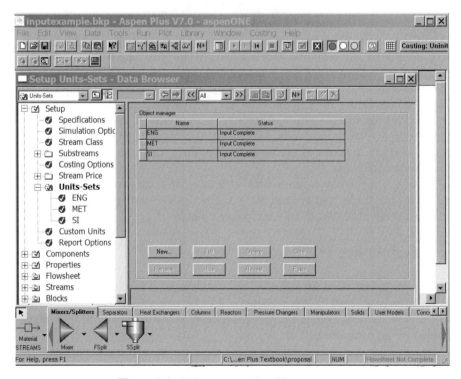

Figure 1.6   Units-sets and the object manager.

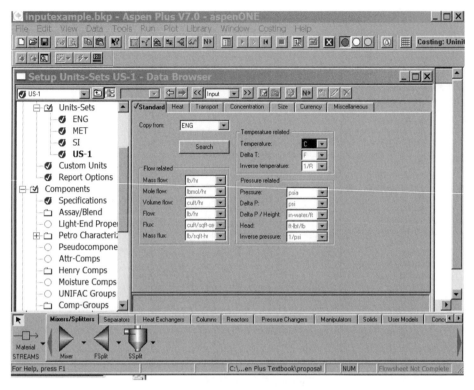

Figure 1.7 New units.

The Copy from entry permits selection of which unit set to use as a base. Note the selection of degrees C from the temperature drop-down list. Each unit's entry has an appropriate drop-down list, which will enable a user to customize units.

Custom units for a specific variable—for example, special composition units—may be defined by selecting the Custom Units entry under the Setup list.

Aspen Plus input forms, displays, and reports are generated using the units selected. Input displays typically have a drop-down list adjacent to input boxes which permits the user to select the units of the input data required. This does not, however, affect the units used for output.

Selection of the Report Options entry under the Setup list permits customization of the Aspen Plus results displays and the. txt report, which can optionally be generated when results are available. Details are given in Section 1.11.

1.7 COMPONENTS

Selecting the Components item on the Data Browser panel produces Figure 1.8. Components may be entered by either component name or chemical formula. Aspen Plus also requires that a nickname, Component ID, which is used in all stream reports, be provided for all components.

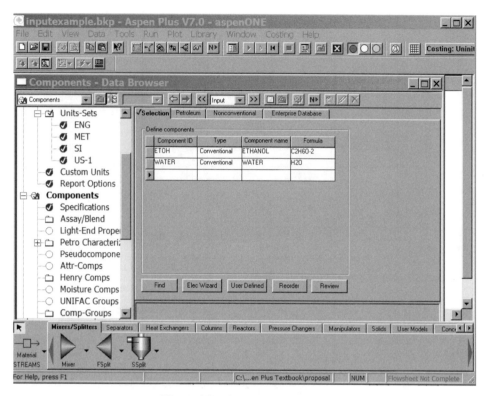

Figure 1.8   Component selection.

An entry in the box Component ID is a user-provided component short name employed by Aspen Plus for report purposes and in some cases, such as water, is recognized as the component water. An entry is always required. Alternatively, the user may enter a proper component name or component formula. If neither is recognized as an entry in the database, the user may select the Find button and Aspen Plus will display a set of names or formulas that incorporate the entry. For example, entering the formula C_7H_8 gives the results shown in Figure 1.9. Upon selection of the component of interest, pressing the Add selected compounds button enters the component into the data associated with the current Aspen Plus run.

If a component does not exist in the Aspen Plus database, choosing User Defined in Figure 1.8 produces the User Defined Component Wizard shown in Figure 1.10.

Note that several properties are required. After entering the Component ID and formula, the Basic Data and Molecular Structure displays shown in Figure 1.11 appear. Known experimental data and structural information are entered. As the data-entry process proceeds, an option for Aspen Plus to estimate any missing data appears. Chapter Two concerns methods for estimating data which may be employed selectively for supplying the missing data above rather than permitting Aspen Plus to provide missing data using default methods.

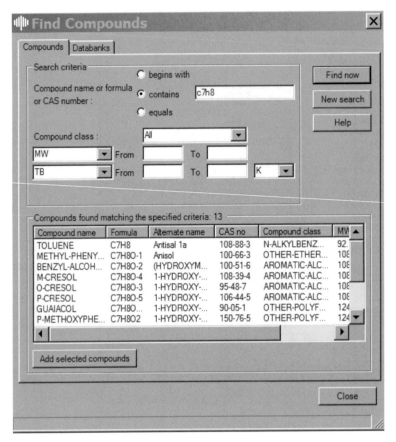

Figure 1.9 Component search.

1.8 PROPERTIES

The first display when selecting properties for a simulation is given in Figure 1.12, where the choice of a global property method that applies to an entire flowsheet may be selected. If, however, the simulation involves a situation where more than property method is required, for example, a process involving distillation (vapor–liquid equilibrium) and extraction (liquid–liquid equilibrium), each block in the flowsheet is identified as part of a unique flowsheet section when the flowsheet is created, and is identified with a particular property method by choosing the Flowsheet Sections tab. A display similar to Figure 1.12 appears; the properties method for each flowsheet section can be identified and a global property method is not used. The Tools menu contains the Property Method Selection Assistant, which can be used as a guide for selection of a property method for specific applications; alternatively, the small box to the left of the Uniquac selection in this example, although not explicitly labeled, performs the same function.

For this example process, sections S-1 and S-2 are assigned the property methods Uniquac and Uniq-2. Aspen Plus sets up the required Uniquac binary parameters for each selection. Care must be taken that the appropriate database is accessed for each set

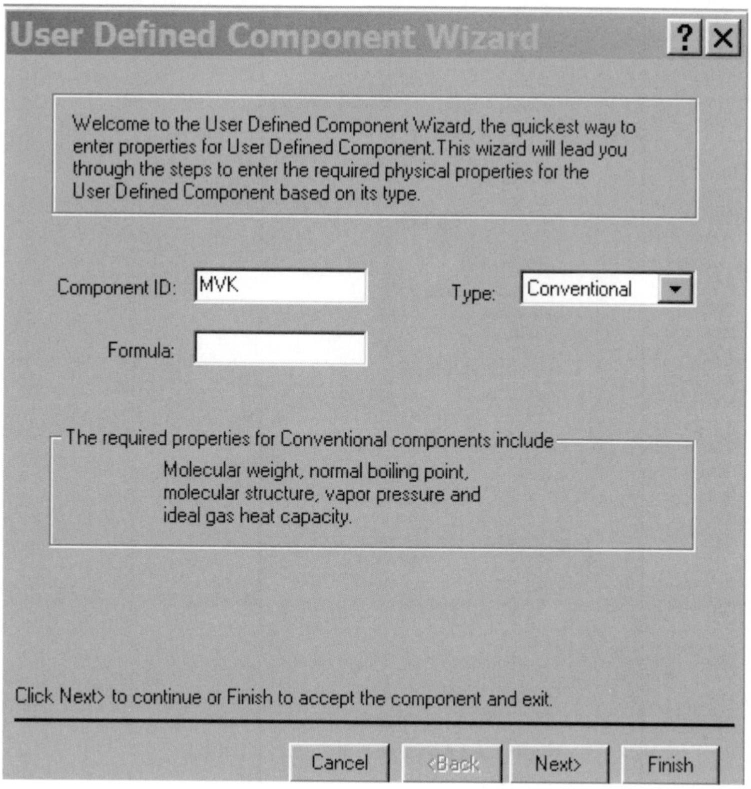

Figure 1.10   User defined component.

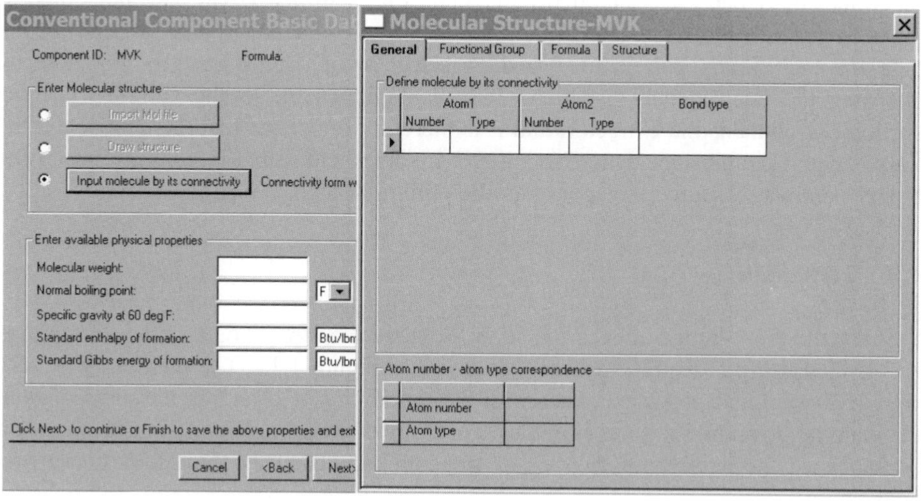

Figure 1.11   User defined component data and molecular structure.

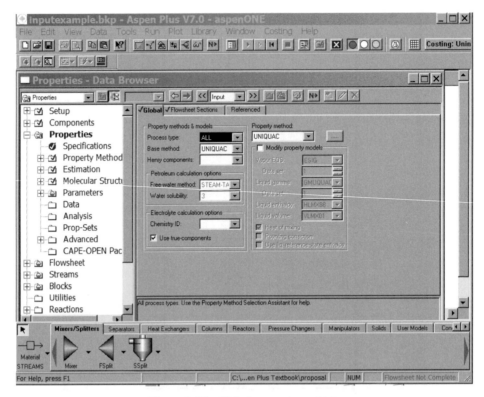

Figure 1.12 Global property method.

of parameters. As an example, if the Data Browser topic Properties/Parameters/Binary Parameters is selected and the listing Uniq-1 is clicked, Figure 1.13 appears. This shows the values of Aspen Plus–supplied parameters for each binary pair and the source of the data used in the regression. In Figure 1.13, note that the source of equilibrium data for each binary pair is identified. A selection of data sources is available by choosing the tab Databanks, which produces Figure 1.14. In the current example, section S-1 contains the extractor, and it is therefore important that the data source be liquid–liquid equilibrium data. Section S-2 involves a distillation column; hence the binary parameters source is vapor–liquid equilibrium data.

1.9 STREAMS

Pushing the Next button moves the input sequence to stream data. All feed streams are defined using a display such as that given in Figure 1.15. The data entry is very straightforward and provides a user with the possibility of changing the units of both the material flow and the state variables. For a stream of a single component or several components, the number of degrees of freedom calculated by the Gibbs phase rule applies:

$$F = C - P + 2$$

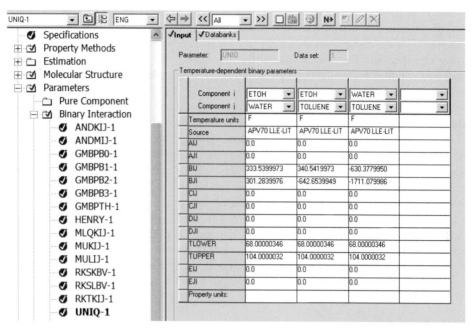

Figure 1.13   UNIQ-1 parameters from Aspen Plus databank.

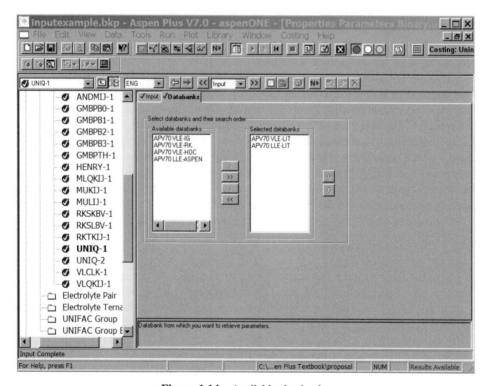

Figure 1.14   Available databanks.

where F is the degrees of freedom, C the number of components, and P the number of phases. For a single-phase system, $F = 2$; therefore, two specifications are required. For an n-component system, $F = n + 1$. Since $n - 1$ mole fractions need to be specified and the last calculated by the sum of the mole fractions equal to 1, only two additional specifications are required. In both cases these are usually, but not necessarily, temperature and pressure.

When two phases in equilibrium are involved, only one degree of freedom is available. For example, for a one-component stream such as a saturated liquid, specification of the temperature fixes the (vapor) pressure. But in such circumstances it is necessary to state the fraction of the mixed stream that is vapor (or liquid). For a saturated liquid the V/F specification would be a very small number, such as 0.00001. For a multicomponent stream the situation is identical and it is necessary to specify either temperature or pressure and the vapor or liquid fraction.

If the process contains tear (recycle) streams, they will not be treated as required input, and pushing the Next button may not suffice. Typically, Aspen Plus will assign zeros as starting values to the variables that are to be converged, but if a user wishes to provide starting values, the stream name under the streams list in the Data Browser can be clicked and a display analogous to Figure 1.15 will appear.

1.10 BLOCKS

When all stream input has been completed, pushing the Next button will result in the appearance of the first input form for a block that requires data. Details of block input are addressed in other chapters. After the data input forms for the first block are completed, pushing the Next button will produce the forms for the next block in the process until all the block data have been entered.

The browser list in Figure 1.15 shows some additional topics that may be appropriate for a user's simulation. These will not appear when the Next button is pushed; however, they may be selected by clicking on the subject. For example, clicking the Convergence entry permits the selection of the convergence method and parametric default values. These subjects are addressed in other chapters. When data entry for all blocks and supplementary data entry is completed, selecting the Next button produces a dialogue to enable execution. All input can be reviewed, prior or after execution, by selecting the $\ll$ or $\gg$ keys above the Flash Options tab near the top of Figure 1.15 and selecting input from the drop-down list between the double arrows.

1.11 VIEWING RESULTS

In preparation for viewing the results, prior to execution of a simulation, a user may customize reports by selecting Report Options, as shown in Figure 1.16. Customization is available under each of the tabs. The stream tab illustrates the variety of options available for selecting the stream information to be displayed.

At the conclusion of a simulation, the run control panel is usually displayed. If not, it can be viewed from the drop-down list under the main menu item View. This contains a brief summary of the execution of each block and a list of error messages. It is important that users check and correct any errors that might have occurred. When a process contains tear streams, errors that occur during the iterations will be presented.

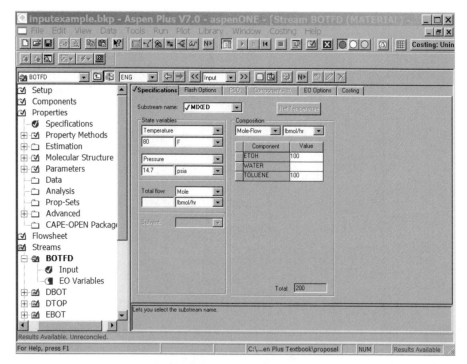

Figure 1.15   Stream specifications.

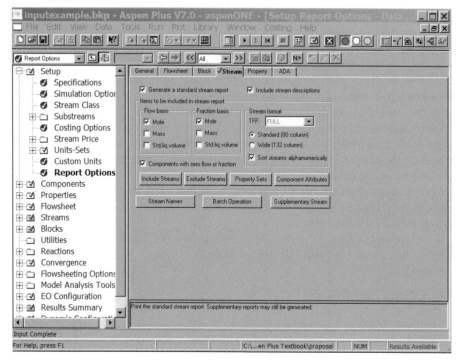

Figure 1.16   Report options.

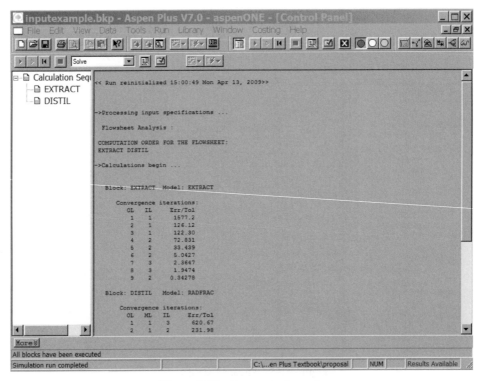

Figure 1.17 Run control panel.

It is not uncommon that errors occur before convergence, but when the process has converged, there should be none. Figure 1.17 displays an example of the run control panel. The results of the simulation can be viewed by selecting the Results button, which contains a check symbol overlaid on a file symbol on the fourth row from the top of the display.

Selection of the Results button produces a display similar to Figure 1.15, with the central drop-down list displaying results. Selection of the ≪ or ≫ keys permits paging through all the results. If All is selected from the drop-down list, one may view both input and results in order.

When a simulation has been completed, selection of Report from the main menu item View will produce a. txt report of the complete simulation, in a Notepad window that can be copied and pasted.

The file Chapter One Examples/inputexample.bkp was used to produce all of the figures in this chapter as well as the inputexample.txt file.

1.12 OBJECT MANAGER

In certain data input situations there are specialized setups that involve the specification of material that requires a special set of dialogues to define what is required. Examples

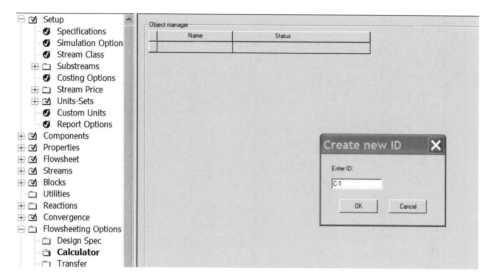

Figure 1.18 Object manager.

are: definition of a set of properties to be displayed, regression of a data set to a thermodynamic model, and specification of the value of a block output variable. This type of input is managed through use of the Object Manager. These situations typically begin with the identification of a function, such as a regression, followed by a series of choices and specifications. An example of the use of the Object Manager for a Calculator block, which is used for in-line calculations, is given in Figure 1.18. The user creates an ID such as C-1, which is followed by all the input forms required, including association of calculation variables to flowsheet variables and definition, in Fortran, of the calculation details.

1.13 PLOTTING RESULTS

Aspen Plus has a plotting facility accessed through the Plot menu. The facility either provides the selection of independent and dependent variables manually or by means of a Plot Wizard, which permits the user to select a preconfigured plot. As an example, after obtaining a converged solution to a distillation column, selecting the Plot Wizard, and pushing the Next button, a collection of distillation-oriented plots, shown in Figure 1.19 is available. Selecting the Comp plot produces Figure 1.20, which shows the change in vapor composition through the column. For situations where preconfigured plots are not available but a display of tabular output is available, such as the results of a sensitivity analysis (details in Chapter Five), one may select a column of data and assign it to either the x or y coordinate. It is also possible to assign two variables to the y coordinate by holding down the control key when selecting the columns of data that are to be employed. If the scale or title of an axis or title of a plot is

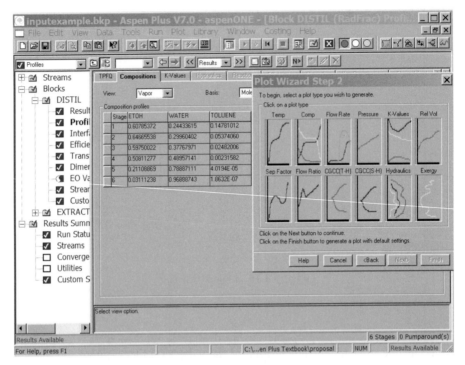

Figure 1.19 Plot wizard.

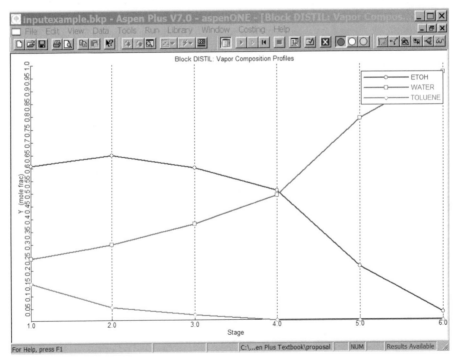

Figure 1.20 Sample plot.

not suitable, one may select it by clicking, and a display that offers editing options appears.

REFERENCES

Aspen Plus version 7.0 documentation.
Monsanto Corporation, *Flowtran Simulation: An Introduction,* 1974.

CHAPTER TWO

PROPERTIES

Aspen Plus offers two possibilities for accessing properties: Property Analysis and Property Estimation. Each is invoked from the drop-down, Run-type menu (Figure 2.1). The Property Analysis function can display pure component values such as the critical compressibility factor, temperature-dependent properties such as the ideal gas heat capacity, and mixture properties from a variety of data banks and mixture properties from multicomponent functions such as equations of state. The Property Estimation capability can be used to estimate virtually the same values as are stored in the Aspen Plus database for user-defined components. References in this chapter to pure component properties, equations of state, activity coefficient equations, and property equations can all be found in the Aspen Plus Physical Property System documentation.

2.1 PURE COMPONENT DATA BANKS

All of the many data banks available in the Aspen Physical Property System can be identified by clicking the Help button at the top of Figure 2.1. The primary database is Pure22. To see the details of its content, searching Help for Pure22 will present various alternatives, one being to display the Pure22 Databank. The list below, taken from Aspen Plus's documentation, describes the property categories for which parameters are stored.

- Universal constants, such as critical temperature and critical pressure
- Temperature and property of transition, such as boiling point and triple point
- Reference-state properties, such as enthalpy and Gibbs free energy of formation
- Coefficients for temperature-dependent thermodynamic properties, such as liquid–vapor pressure

---

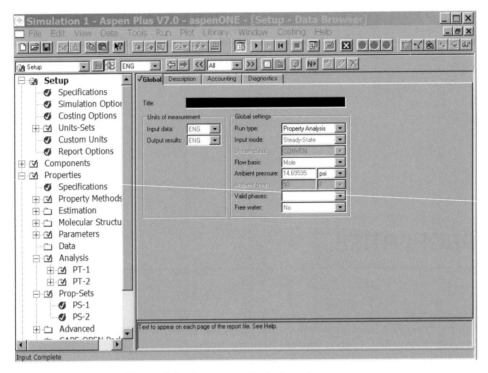

Figure 2.1 Property analysis drop-down menu.

- Coefficients for temperature-dependent transport properties, such as liquid viscosity
- Safety properties, such as flash point and flammability limits
- Functional group information for all Unifac models
- Parameters for Soave–Redlich–Kwong and Peng–Robinson equations of state
- Petroleum-related properties, such as API gravity, octane numbers, aromatic content, hydrogen content, and sulfur content
- Other model-specific parameters, such as the Rackett and Uniquac parameters

The list of components stored can be found by clicking the link Pure Component Databanks and at that link, clicking the Pure Component Databank Parameters, which provides the parameter names, descriptions, units, and availability in other Aspen Plus data banks. An example of such data is shown in Figure 2.2. This display is invoked by selecting the Run type Property Analysis, entering components (in this example, methanol and water), and selecting Tools, Retrieve Parameter Results, and OK. The display will be found under Results, Pure Component. Pure component temperature-dependent results are also available from the T-Dependent tab; for example, the selection of parameter Plxant-1 refers to the coefficients of the extended Antoine equation, as shown in Figure 2.3. The extended Antoine equation is

$$\ln p^* = C_1 + \frac{C_2}{T + C_3} + C_4 T + C_5 \ln T + C_6 T^{C_7} \qquad C_8 \le T \le C_9 \qquad (2.1)$$

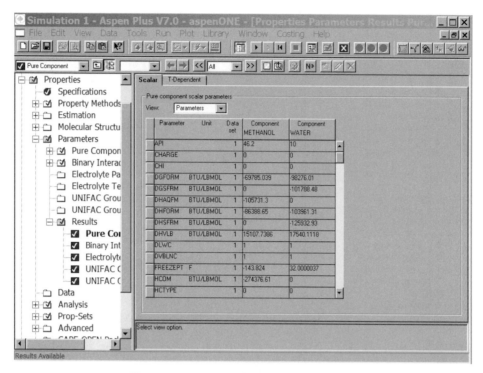

Figure 2.2   Pure component scalar properties.

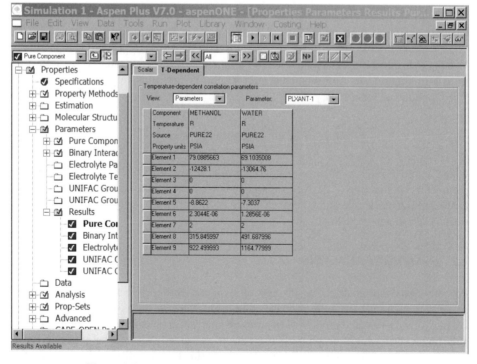

Figure 2.3   Pure component temperature-dependent properties.

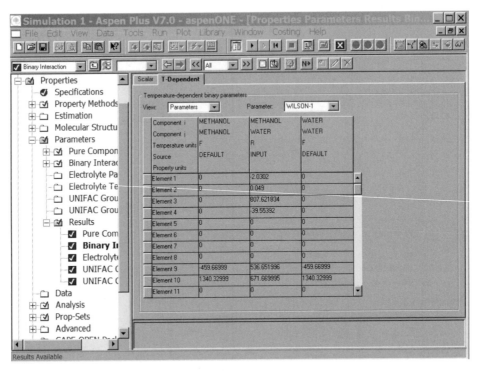

Figure 2.4   Temperature-dependent binary interaction parameters.

where the element numbers in Figure 2.3 correspond to the parameters of equation (2.1).

Selection of the Results Binary Interactions with the tab T-Dependent and parameter Wilson-1 displays the binary interaction parameters of the methanol–water system as shown in Figure 2.4. The elements refer to the Aspen Plus implementation of the Wilson equation for a binary pair:

$$\ln \gamma_{ij} = a_{ij} + \frac{b_{ij}}{T} + c_{ij} \ln T + d_{ij} T + \frac{e_{ij}}{T} \qquad (2.2)$$

where i and j refer to the components and the element numbers correspond to the parameters of equation (2.2); for example a_{ij} is element 1, and elements 9 and 10 refer to the applicable temperature range. Similar results appear when selecting a different activity coefficient equation.

2.2 PROPERTY ANALYSIS

Property Analysis enables the display of temperature-dependent properties at user-selected temperatures as well as phase equilibrium data for both two and three phases. Display of user-selected properties requires an analysis to be defined, including the variable range over which the properties are to be presented. Aspen Plus presents forms that need to be completed, similar to flowsheet run forms, followed by a properties specification input. Clicking the Next button initiates a display in which the property

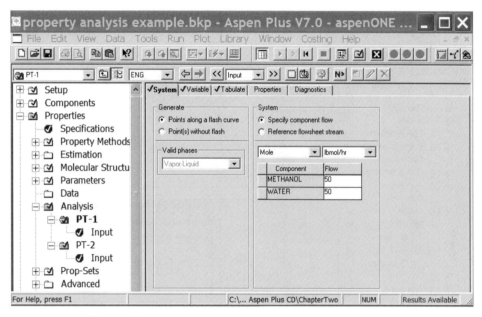

Figure 2.5   Example of properties analysis specification PT-1.

analysis to be generated is specified. For example, a vapor–liquid equilibrium properties specification, labeled PT-1, is shown in Figure 2.5, where points along a flash
curve are chosen in anticipation of preparing a $T-x-y$ diagram. Aspen Plus requires
the specification of composition, even though it is not needed for this example. Note
the appearance of the adjusted variable range display shown in Figure 2.6. Prior to
completing the analysis input, Aspen Plus prompts the user to define which properties are required. These are defined in a Prop-Set, with an example labeled PS-1, and
displayed in Figure 2.7, where the properties required are the bubble and dew points
at 14.7 psi as a function of composition. It is possible for several property analyses
and property sets to be executed with one Aspen Plus run, and a display is provided
to associate a Prop-Set with an analysis. An example of such an association is shown
in Figure 2.8. The results are the tabular display shown in Figure 2.9, which can be
plotted with Aspen Plus's Plot function, as shown in Figure 2.10.

Another example of a Prop-Set, PT-2, which deals with an analysis of temperature-
dependent properties, is shown in Figure 2.11. The tab Qualifiers is used to define the
components, the phases, and other specifications for which properties are desired with
the associated Prop-Set PS-2. Tabular results are shown in Figure 2.12. A property
analysis example may be found at Chapter Two Examples/property analysis example.

2.3   PROPERTY ESTIMATION

There are many methods for estimating pure component properties such as the critical
pressure. When using Aspen Plus to estimate such values, a user is offered a choice
of methods, most of which are described by Poling et al. (2000), and in some cases,
in an earlier edition of the same book. As an example, Joback's method for estimating

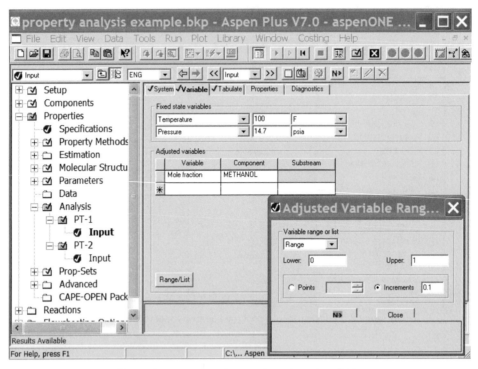

Figure 2.6 Range definition for adjusted variable.

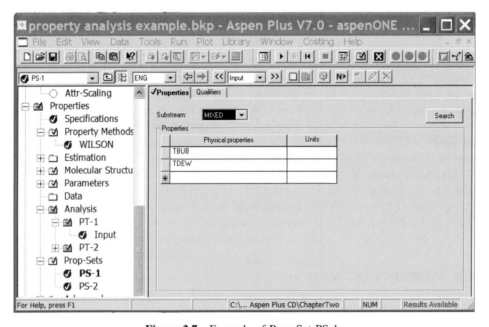

Figure 2.7 Example of Prop-Set PS-1.

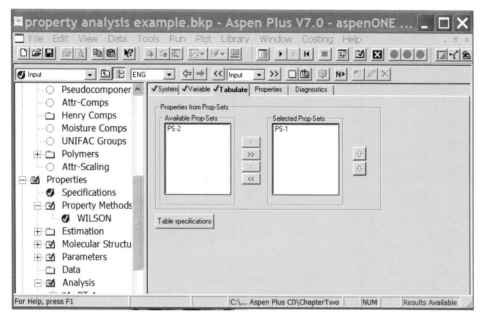

Figure 2.8   Association of PS-1 with PT-1.

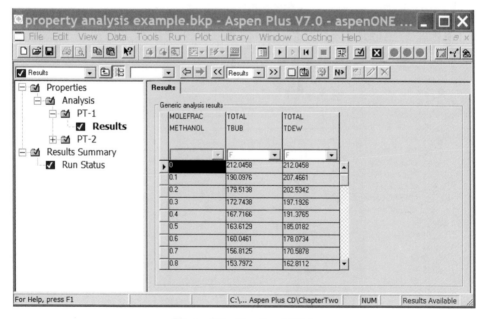

Figure 2.9   Results of PT-1.

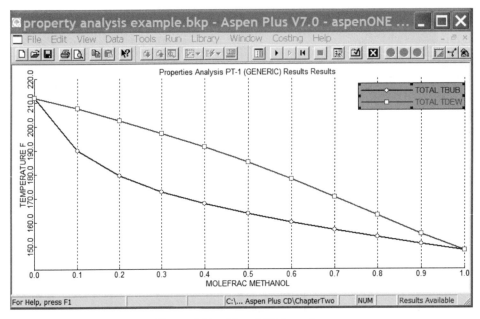

Figure 2.10 Plot of PT-1 results.

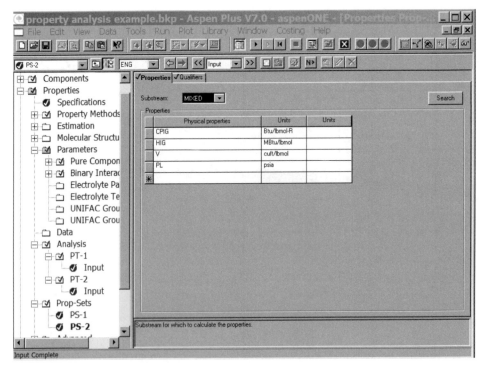

Figure 2.11 Variable selection for Prop-Set PS-2.

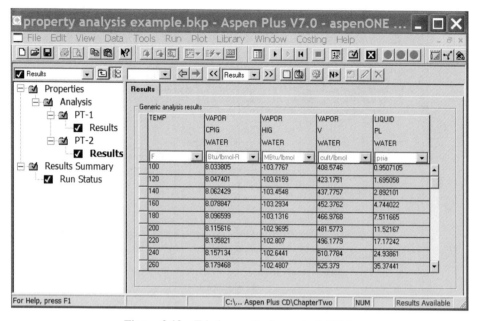

Figure 2.12 Tabular results for analysis PT-2.

the critical temperature, is given by

$$T_c(K) = T_b \left\{ 0.584 + 0.965 \left[\sum_k N_k(tck) \right] - \left[\sum_k N_k(tck) \right]^2 \right\}^{-1} \qquad (2.3)$$

where *tck* are structural group contribution values and N_k are the number of appearances of that group in the molecule. Note that the critical temperature is directly proportional to the boiling point modified by group contribution functions.

Equations of state provide the means for estimating many other properties, such as the effect of pressure on enthalpy at constant pressure. Since the equation-of-state parameters are composed primarily of critical properties and in many cases the acentric factor (derived from vapor pressure), experimental values for the boiling point and vapor pressure data are very important.

The property constant estimations system within Aspen Plus, sometimes referred to as PCES, is initiated by the same methods as in Chapter One, except that the Run type Property Estimation is entered. This function is used when a user-defined component is required. The component *should not* be entered into the component list, but rather, the new component wizard, shown in Figure 2.13, should be employed. Aspen Plus prompts for any available basic data and molecular structure, as shown in Figure 2.14 followed by Figure 2.15 for additional data. It is imperative that the radio button to estimate properties by the Aspen Plus system be selected. Methyl vinyl ketone (MVK) is an example of a user-defined component. The molecule's structure is

$$CH_3-CO-CH=CH_2$$

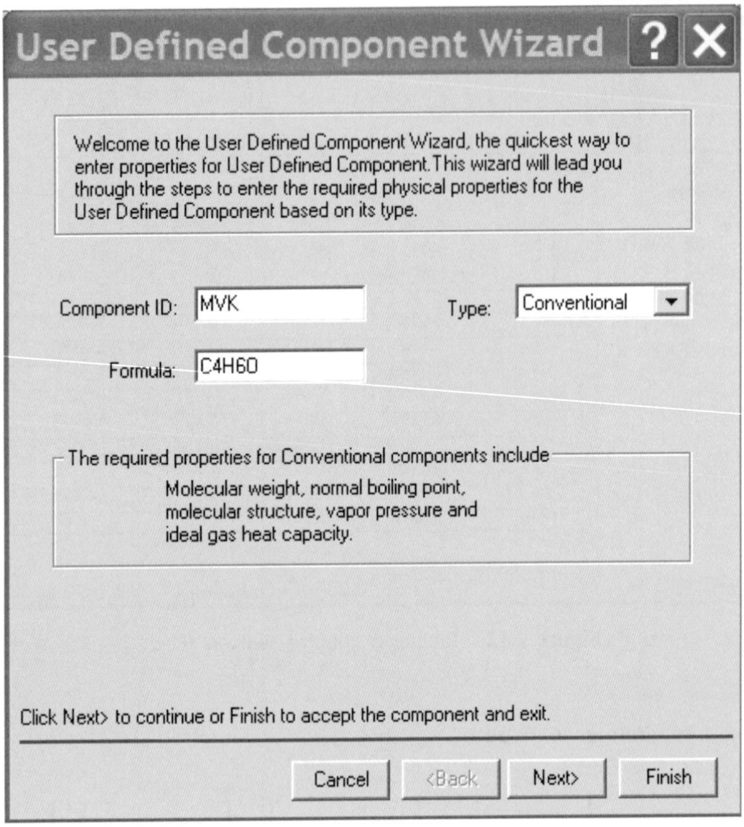

Figure 2.13 User defined component wizard.

The molecule's structure, defined in Figure 2.14, is entered by assigning a number to each nonhydrogen element and defining the bond type between them. In this case the carbons are numbered from the left 1, 2, 4, and 5, and the oxygen is numbered 2.

For a multicomponent system, components stored in an Aspen Plus database or additional user-defined components may be added. Pushing the Next key presents Figure 2.16, which enables a choice of estimation options. If the choice to, estimate all missing parameters is made, it is possible that only the parachor will be estimated. It is preferable to choose the radio button and estimate only the parameters selected, in which case check boxes for selection of parameter types become available and each property desired can be selected via the parameter and/or property lists in the tabbed categories. Figure 2.17 shows an example in which all estimation methods have been selected. This may generate a warning message for some methods, due to their limitations on execution. The temperature-dependent properties input are similar to pure component properties except that an independent variable range must be specified.

The tab Binary is used to estimate a_{ij} pairs for activity coefficient equation pairs from estimates of infinite dilution activity coefficients by one of the Unifac methods available in Aspen Plus. It is also possible to estimate the k_{ij} parameter for the SRK equation of state. The component list must contain additional entries which specify the components

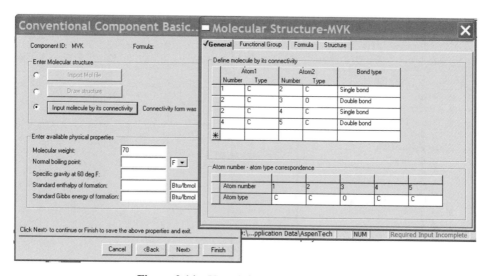

Figure 2.14   User defined component data.

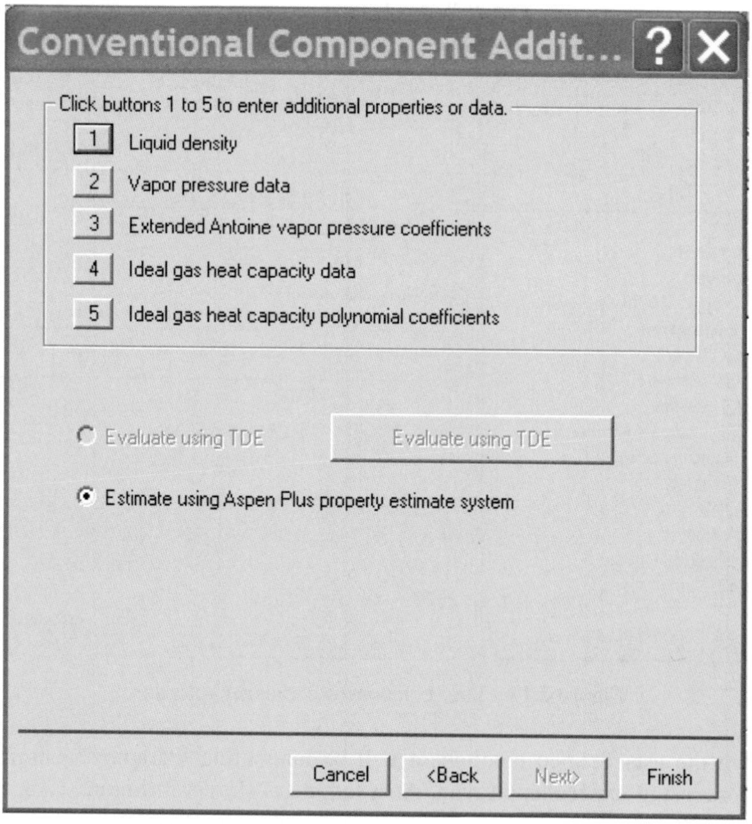

Figure 2.15   Additional user defined component wizard.

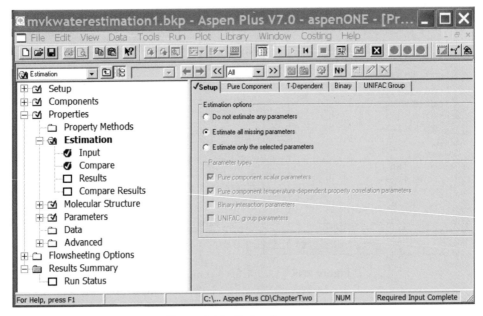

Figure 2.16 Estimation setup.

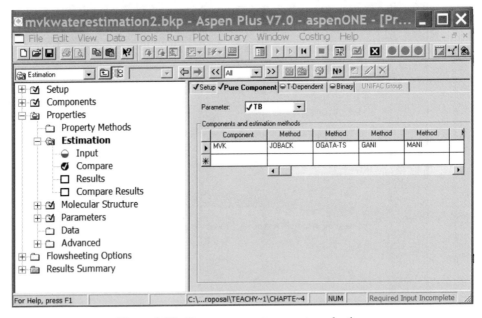

Figure 2.17 Pure component parameter selection.

with which the user-defined component will be interacting. Parameter estimation can also be done using experimental data. An example is shown in Figure 2.18.

Occasionally, it may be necessary to supplement the structural data supplied in Figure 2.14 with group contribution data for one of the methods available. An example of Unifac data for the MVK molecule is given in Figure 2.19. The identification of

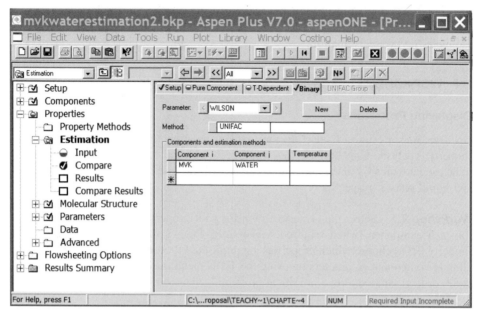

Figure 2.18   Binary parameter estimation setup.

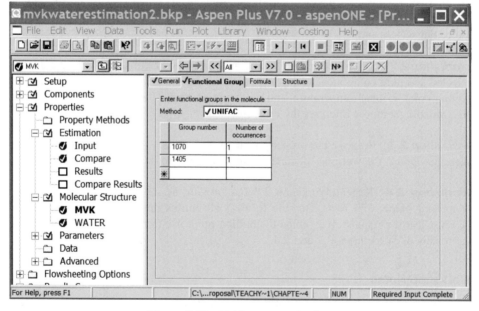

Figure 2.19   Unifac group selection.

the group numbers is provided at the bottom of the display, but they cannot be seen unless the display is full size. The setup and results for this example may be found at Examples/mvkwaterestimation.

2.4   WORKSHOPS

Displaying Properties

Open an Aspen Plus run selecting the blank simulation template with the Run type Property Analysis. Use the Next key to proceed with the data input. The components to be used are mesityl oxide and toluene. Select Unifac and Ideal to represent the vapor and liquid phases, respectively.

Workshop 2.1   Select the Generic option for your property type and enter 50 mol/hr for each component when required. Generate a $T-x-y$ diagram at 14.7 psi. Use a Property Set to choose which properties are required. Plot the $T-x-y$ diagram. Display all pure component parameters by using the Tools menu. Save the values of the boiling point, critical temperature, and critical pressure.

Estimating Properties

Open an Aspen Plus run by selecting the blank simulation template with the Run type Property Analysis. Use the Next key to proceed with the data input. The components to be used are mso-e (mesityl oxide) and toluene. Treat mso-e as a user-defined component even though it is in the data bank. Provide the molecular structure and molecular weight only.

Workshop 2.2   Select Setup and estimation options, and estimate only the parameters selected and pure component scalar properties. Under the Pure Component tab, choose boiling point, critical temperature, and critical pressure for the component mso-e using all available methods. Interpret the error messages.

Workshop 2.3   Remove the property estimation methods that caused the error message, but ignore the warning message. Reinitialize and rerun.

Workshop 2.4   Repeat Workshop 2.3 but provide as input the true boiling point of mesityl oxide: 129.76°C. Be sure to clear all entries under the setup menu entry PCES. Clear the request to estimate the boiling point. Reinitialize and rerun. Compare the results of Workshops 2.3 and 2.4.

Workshop Notes

Workshop 2.1   Unifac is used to estimate the vapor–liquid equilibrium for this system to demonstrate the possibility of obtaining such data in the absence of experimental values. If possible, experimental verification is extremely important.

Workshop 2.2   Although all methods for estimating the three properties appear available, some of the methods have limitations. For example, Ogata's method for estimating boiling points is limited to one radical per molecule. Some of these difficulties inhibit execution of the Property Estimation run.

Workshop 2.3   Upon removal of those methods with problems from the property specifications, the execution is satisfactory despite the warning message concerning the boiling point and vapor pressure. Note that the two methods for estimating the boiling point are about 10 K apart. Notice also by comparison to the database values presented in Workshop 2.1 that the estimated boiling points by both methods are off by 5 to 15°F.

Workshop 2.4   Comparing the critical temperatures predicted by Workshops 2.3 and 2.4, where the experimental boiling point is included in the data, the Workshop 2.3 data average about 589 K, and the Workshop 2.4 data, 597 K. The critical pressure does not depend on the boiling point.

REFERENCES

Aspen Plus version 7, Physical Properties System documentation.

Poling, B. E., Prausnitz J. M., and O'Connell, J. P., *The Properties of Gases and Liquids*, 5th ed., McGraw-Hill, New York, 2000, Chaps. 2 and 3.

CHAPTER THREE

THE SIMPLE BLOCKS

Aspen Plus provides several blocks that are suitable for, but not limited to, material-balance-only calculations: the stream splitter, the mixer, and the simple separator blocks. Additionally, the stoichiometric reactor, covered in Chapter Ten, is also very useful for material-balance-only calculations. These blocks are especially important since when they are combined to form a simple process model, the results are useful as an initial estimate for a rigorous model of a process. In some cases, such calculations are used as the basis for design.

Aspen Plus also provides two blocks with general utility in various aspects of simulation. The stream duplicator block enables, for example, the execution of duplicate processes with different thermodynamics and parameters but the same feeds. The stream multiplier enables all elements of an input stream to be modified by a calculated multiplier subject to the specification of a block output, such as the composition of a component in an output stream.

3.1 MIXER/SPLITTER BLOCKS

The blocks Mixer, Fsplit, and Ssplit are available from the Aspen Plus model library shown on the bottom of the Aspen Plus displays under the Mixers/Splitters tab, as shown in Figure 3.1.

3.1.1 Mixer Block

The Mixer block combines an arbitrary number of input streams and produces a single stream by a simple material balance. If the "no heat balance" option is chosen, no other input is required. An example is shown in Figure 3.2. If F_i and h_i represent the

Teach Yourself the Basics of Aspen Plus™   By Ralph Schefflan
Copyright © 2011 John Wiley & Sons, Inc.

Figure 3.1 Mixers/splitters.

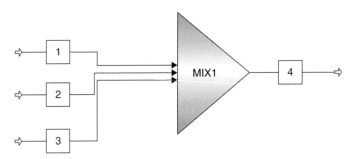

Figure 3.2 Mixer block.

flow rate and enthalpy/mole of stream i, the material balance and energy balance are

$$F_4 = F_1 + F_1 + F_3 \tag{3.1}$$

$$h_4 F_4 = h_1 F_1 + h_2 F_2 + h_3 F_3 \tag{3.2}$$

When energy balances are selected, it is necessary to specify the outlet stream's pressure or the pressure drop across the block. If no pressure is specified, the block uses the lowest pressure of the feed streams. Additionally, the allowable phases of the output stream must also be specified. Since the output stream is of known composition and pressure and its molar enthalpy is calculated using equation (3.2), its state is defined, and therefore an adiabatic flash yields its temperature.

3.1.2 Fsplit Block

The Fsplit block is designed to divide a single stream or a mixture of streams of known state (i.e., flow rate, composition temperature, and pressure) into an arbitrary number of product streams with identical states. Note that since the state is known, the enthalpy/mole is calculable. A choice of product stream split fraction (based on the sum of the flow rates of the feed streams), mass or molar flow rate, volumetric flow rate, or flow rate of a component may be made. If there are n product streams, specifications for any $n - 1$ streams must be made. Figure 3.3 shows an example of an Fsplit block.

When the model is executed, all the feed streams are mixed and the combined flow rate, composition, and molar enthalpy is computed. It is important to note that any of a block's product specifications can be converted to product stream split fractions. This provides the basis for the describing equations of the model. For example, if the molar flow rate of component i in stream j, p_i^j, is specified, the split fraction α_i can

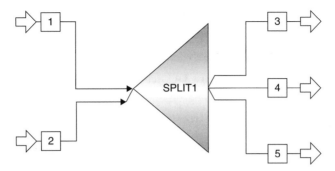

Figure 3.3   Fsplit block.

be calculated as

$$\alpha_i = \frac{p_i^j}{f_i} \tag{3.3}$$

where f_i is the combined feeds flow of component i. Componential material balances are as follows. Let F and h represent the total flow rate and the molar enthalpy of the combined feed streams, and let P_j and α_i represent the flow rate and split fraction of the product streams. Since there are n product streams $n-1$ split-fraction material balance equations, each described by the following equation, may be written for the product streams:

$$P_i = \alpha_i F \tag{3.4}$$

with the flow rate of the last product stream calculated by the overall material balance

$$P_n = F - \sum_{k=1}^{n-1} P_k \tag{3.5}$$

The combined feed molar enthalpy, h, is assigned to each product stream. The pressure of all the product streams is identical and is either specified on the flash tab or if not specified, the lowest input stream pressure is used and thus all product streams are at the same state. If the energy balance option is used for the simulation, the temperature of a product stream is calculated by an adiabatic flash.

The Ssplit block available from the model library, except for the input format, has the same functionality as Fsplit.

3.2 SIMPLE SEPARATOR BLOCKS

The simple separator blocks Sep and Sep2 are available from the Aspen Plus model library shown on the bottom of the Aspen Plus displays under the Separators tab, as shown in Figure 3.4.

Figure 3.4 Separators.

3.2.1 Sep Block

When the Sep model is executed, all the feed streams are mixed and the combined flow rate, composition, and molar enthalpy is computed. The block permits the assignment of the flow rates or componential split fraction of each component in each of the $n - 1$ product streams, based on the combined feeds. The composition of the nth stream is calculated by the overall material balance. An example of a Sep block is shown in Figure 3.5.

As part of the block's input specifications, there is a tab, Outlet Flash, which enables a user to specify the state of each outlet stream. A flash of each stream establishes the molar enthalpies and facilitates calculation of the block energy balance.

Since Sep is a typical sequential modular block, for an n-component, m-product stream configuration, with the temperature and pressure of the feeds and products specified, a degree-of-freedom analysis shows that for a material-balance-only application there are nm unknowns. If $n(m-1)$ products are associated with a specification such as the fraction of the sum of the feeds of a component, to exit in a particular stream, n unknowns remain to be calculated by a material balance; thus, the composition is calculable for all streams. For an energy balance calculation there is one additional unknown per product stream: the molar enthalpy. Since the composition temperature and pressure of each stream is known at the conclusion of the material balance calculations, an adiabatic flash produces the molar enthalpy of all streams, and an overall energy balance can be performed.

If f_i^k represents the quantity of component i in feed stream k, the amount of component i entering the block, F_i, is given by

$$F_i = \sum_{k=1}^{l} f_i^k \tag{3.6}$$

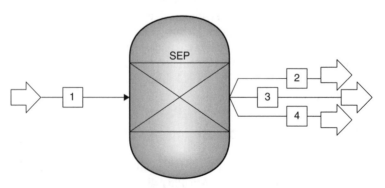

Figure 3.5 Sep block.

where l is the number of feed streams. If the split fraction of a particular component i, in stream j, α_i^j, is given, the quantity of component i leaving the block in stream j, S_i^j, is given by

$$S_i^j = \alpha_i^j F_i \tag{3.7}$$

Alternatively, the quantity of component i leaving the block in stream j, S_i^j, can be specified directly. These variables in the equations do not depend on each other, and thus the solution, where not uniquely specified as in equation (3.7), is direct.

3.2.2 Sep2 Block

The Sep2 block is designed to model a simple two-product separator such as a distillation or extraction process. Like the Sep model, when a Sep2 block is executed, all the feed streams are mixed and the combined flow rate, composition, and molar enthalpy is computed. For n product streams, the block permits assignment of the flow rates or componential split fraction of each component in each product stream of $n - 1$ streams, based on the combined feeds. The Sep2 model, however, provides more flexibility in specification than that of the Sep block. The additional specs allowed are overall stream split, that is, the ratio of flow of a product stream to the sum of the flows of the feed streams, and the mole or mass fraction of a component in a product stream. The composition of the nth stream is calculated by the overall material balance. An example of a Sep2 block is shown in Figure 3.6.

Unlike the Sep block, there is interdependency between some of the block specifications. Equations (3.6) and (3.7) describe the specifications of the Sep block that apply to the Sep2 block, but the other Sep2 specifications involve interaction between the

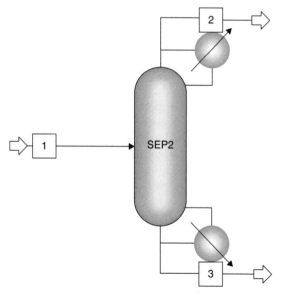

Figure 3.6 Sep2 block.

variables. The mole fraction specification is an example. If x_i^j represents the specified mole fraction of component i in product stream j, when l is the number of components, then

$$x_i^j = \frac{f_i^j}{\sum_{k=1}^{l} f_k^j} \qquad (3.8)$$

applies. An equation similar to (3.8) which describes the total flow specification is

$$\beta_j = \frac{\sum_{k=1}^{l} f_k^j}{\sum_{m=1}^{p} F_m} \qquad (3.9)$$

where β_j is the overall split fraction of stream j to the sum of flows of all p feeds, F_m represents the sum of the flows of all components in feed stream m, and l represents the number of components.

Note that the flows of the components may not be known a priori as in the Sep block; therefore, the solution of a set of simultaneous equations in which the unknowns are the individual component flows of product streams is required.

Like the Sep block, there is a tab, Outlet Flash, which enables a user to specify the state of each outlet stream. A flash of each stream establishes the molar enthalpies and facilitates calculation of the block energy balance.

3.3 SOME MANIPULATOR BLOCKS

Manipulators may be found in the Aspen Plus library under the tab Manipulators (see Figure 3.7). The only manipulators considered here are the stream duplicator and stream multiplier blocks.

3.3.1 Dupl Block

An example illustrating the solution of the same problem with both the Sep and Sep2 blocks is shown in Figure 3.8. Upon execution, streams 2 and 3 are identical to stream 1.

A feed of 100 lbmol/hr of methanol, 200 lbmol/hr of water, and 300 lbmol/hr of ethanol is to be separated into an overhead product of 100 lbmol/hr with composition 0.95 mole fraction methanol and 0.04 mole fraction ethanol. The column, operating at 14.7 psi, is to be simulated twice, once with a Sep block and once with a Sep2 block. A Duplicator block is to be used to feed both columns. Figures 3.9, 3.10, and 3.11 show the specifications for both separator blocks.

Figure 3.12 shows that identical results using the file Chapter Three Examples/ DuplicatorExample are obtained; however, the use of Sep would not be straightforward if it were in the middle of a process where the feed was not known a priori.

Figure 3.7 Manipulators.

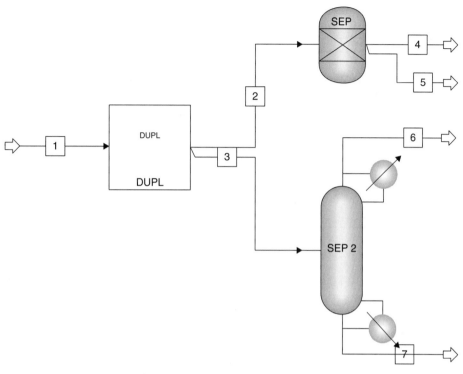

Figure 3.8   Duplicator block example.

Figure 3.9   Sep specifications.

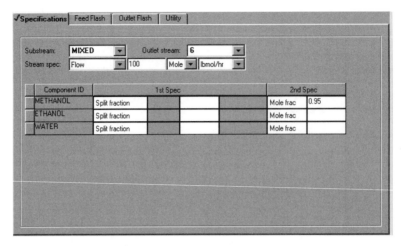

Figure 3.10   Sep2 specifications.

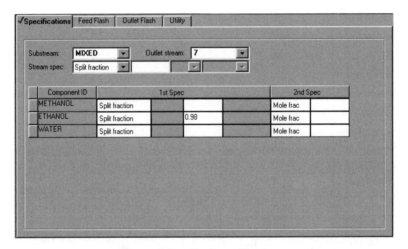

Figure 3.11   Sep2 specifications.

3.3.2 Mult Block

The Multiplier block contains a user-specified parameter which multiplies all the flow rates in the block's feed by a specified value or by a value calculated from another part of the process. An example of the use of the multiplier block is shown in Figure 3.13. In this example, stream 1, the feed to the Mult block, is specified, but stream 2, the feed to the rest of the process, is unknown. In this case a process quantity such as the flow rate of a component in a stream, for example in stream 8, will be targeted to achieve a user-specified value. A design specification, described in detail in Chapter Five, is used to manipulate the value of the multiplier parameter, operating on stream 1, so that stream 2 attains a value such that the desired flow rate of the component in stream 8 is obtained.

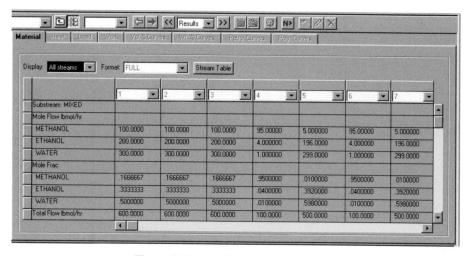

Figure 3.12   Duplicator example results.

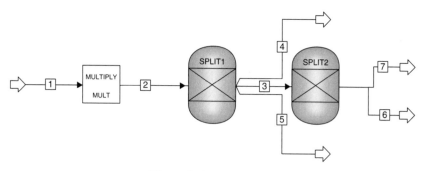

Figure 3.13   Mult block.

3.4   WORKSHOPS

Workshop 3.1a   Three streams are to be mixed. Their specifications are given in Table 3.1. Use the option "material balance only." The property option to be used is Chao-Seader. Solve the flowsheet. The components are propane, n-butane, n-pentane and n-hexane. Stream specifications are given in Table 3.1.

Workshop 3.1b   Repeat Workshop 3.1a using material and energy balances. What do you observe? Is your solution correct?

Workshop 3.2a   Three streams are to be fed to a Fsplit block used to emulate a distillation column with a sidestream. The process specifications are given below. Use the option "material balance and energy balances." There will be three product streams: D, SS, and B. The property option to be used is Chao-Seader. The component list and stream specifications are the same as in Workshop 3.1a. Make the following specifications:

TABLE 3.1   Stream Specifications for Workshop 3.1a

Stream/ Component	Temperature (°F)	Pressure (psi)	Flow Rate (lbmol/hr)	Vapor Fraction
1	200	300		
C3			20	
NC4			30	
NC5			30	
NC6			20	
2	250	400		
C3			30	
NC4			30	
NC5			20	
NC6			20	
3	200			0.5
C3			50	
NC4			0	
NC5			30	
NC6			20	

1. 50% of the combined feeds are to be directed to stream D.
2. 50 lbmol/hr of NC4 will exit in stream SS.

Solve the flowsheet. Is something wrong? What? Why?

Workshop 3.2b   Change Workshop 3.2a such that 20 lbmol/hr of NC4 exits in stream SS. Solve the flowsheet. What do you observe?

Workshop 3.3a   A two-feed, two-product distillation column operating at 1 atmosphere is to be simulated using the Sep2 block. The system components are methanol, water, and normal hexane. You may assume that the system thermodynamics can be represented by an ideal vapor phase and the Unifac correlation for the liquid phase. The componental flow rates of the first feed, a saturated liquid at 14.696 psi, are 50 lbmol/hr of methanol and 100 lbmol/hr of water. The second feed flow rate is 150 lbmol/hr of pure n-hexane at 120°F and 14.696 psi. Both feeds are at 1 atmosphere and are saturated liquids. The column performance specifications are as follows:

1. One-half of the methanol fed is to be recovered in the overhead product.
2. 99% of the water fed is recovered in the overhead product.
3. 99% of the n-hexane fed is recovered in the bottom product.

Set up the problem and solve.

Workshop 3.3b   Specifications 2 and 3 above are to be changed as follows:

2. The mole fraction of water in the overhead product is 0.788.
3. The mole fraction of n-hexane in the bottom product is 0.85.

Modify Workshop 3.3a and solve.

Workshop 3.3c   Specification 2 is to be changed as follows:

2. The mole fraction of water in the overhead product is 0.8.

Modify Workshop 3.3b and solve by hand. What do you observe?

Workshop 3.3d   Specification 2 is to be changed as follows:

2. The mole fraction of water in the overhead product is 0.75.

Modify Workshop 3.3c and solve. What do you observe?

Workshop Notes

Workshop 3.1b   Care must be taken with the specifications of the allowable phases in the product. If there is a possibility that the resulting state is a mixture of phases, it must be specified on the Flash Options/Valid phases data-entry location.

Workshop 3.2a   The specifications stated in the problem have not been achieved. It is recommended that the problem be solved manually. This will show that a negative flow of one of the products results. This problem is specified improperly.

Workshop 3.2c   A satisfactory solution is obtained.

Workshop 3.3a   A satisfactory solution is obtained.

Workshop 3.3b   A satisfactory solution is obtained.

Workshop 3.3c   A satisfactory solution is not obtained. A manual material is given below using the following nomenclature:

f_1, f_2, f_3 are the combined feed flow rates of methanol, water, and n-hexane, respectively.
d_1, d_2, d_3 are the componential distillates.
b_1, b_2, b_3 are the componential bottoms.

The material balance equations are then as follows:

$$d_1 + b_1 = 50$$

$$d_2 + b_2 = 100$$

$$d_3 + b_3 = 150$$

$$\frac{b_3}{b_1 + b_2 + b_3} = 0.85$$

$$\frac{d_2}{d_1 + d_2 + d_3} = 0.8$$

The first specification results in $d_1 = b_1 = 25$. Simultaneous solution of the remaining four equations and substituting the known values for d_1 and b_1 results in the following results: $b_2 = 1.59$; $b_3 = 151.4$; $d_2 = 98.46$; $d_3 = -0.38$. Since there is a negative flow for d_3, n-hexane in the distillate, this is a physically unrealistic solution; therefore, the specifications are improper.

Workshop 3.3d A satisfactory solution is obtained as a result of the change in the overhead specification for water.

CHAPTER FOUR

PROCESSES WITH RECYCLE

Virtually all Aspen Plus blocks are designed to operate as part of a sequential-modular (SM) simulator; that is, given the block's inputs and operating parameters, the block will calculate internal conditions and its outputs. In most cases the block's operating pressure and product stream pressures are specified in some way. This is illustrated by the example of a Flash2 block shown in Figure 4.1.

Streams 1 and 2 are either specified or calculated by a block executed prior to block B1. Stream 5 is a special case of free water, usually zero. The block's products, each of streams 3 and 4, have $n_c + 2$ variables, where n_c is the number of components. These are the flows for each of the components temperature and pressure, for a total of $2n_c + 4$ variables. Counting equations for a Flash2 block, there are n_c material balances, n_c equilibrium equations, and two equations that satisfy the equilibrium requirement that the temperature and pressure of the equilibrium streams, 3 and 4, are equal, for a total of $2n_c + 2$ equations with two degrees of freedom. The specification of any of the possible two flash conditions—for example, the heat added and the vapor/feed ratio—satisfies the block's requirements and permits calculation of the state of both product streams. A process composed solely of SM models simulates the performance of a process.

A variation on such a model poses the question: What temperature should the flash operate at to give a desired product composition, within the constraints of thermodynamic reality? As an example, if flashing a mixture of hydrogen and rocks, the basic laws of nature will prohibit rocks from being in the overhead product. This mode of simulation encompasses a design function. To accomplish this, one may employ a

Teach Yourself the Basics of Aspen Plus™ By Ralph Schefflan
Copyright © 2011 John Wiley & Sons, Inc.

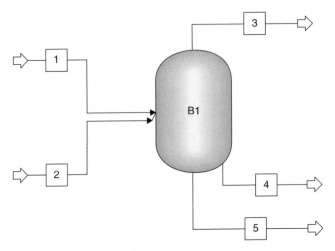

Figure 4.1 Flash2 block.

SM model by releasing a specification on the flash conditions and, instead, specifying the composition desired for one component in a product. This is termed a *design specification*. Details are described in Chapter Five.

4.1 BLOCKS WITH RECYCLE

When blocks are connected for a process with recycle such as Figure 4.2, traditional chemical engineering practice identifies stream 6 as a recycle stream. The technique for solution involves a guess of the component flows, temperature, and pressure of the stream and uses the method of direct iteration. The values of the variables in stream 6 are estimated and the blocks are executed in the sequence React, Flash, and Split,

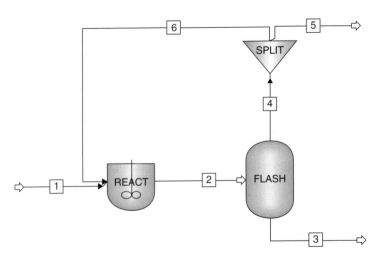

Figure 4.2 Recycle process.

and produce a calculated set of values for stream 6. The values of stream 6 guessed and calculated are compared, and if they are not equal within a set tolerance, the calculated values of stream 6 are used as the set of values for the next iteration in the calculations. The method of direct iteration may require many iterations to converge and does not guarantee a solution. If a particularly difficult process is encountered, it is sometimes beneficial to select the convergence option, direct, and limit calculations to one or two iterations. The maximum number of iterations can be selected under ConvOptions/Methods under the tab corresponding to the convergence method, and chosen as Maximum flowsheet evaluations. This frequently helps in analysis and debugging of models.

The choice of stream 6 as a recycle stream is intuitive, but it is not the only approach. If one were to choose stream 2 as a "forward" recycle, one could still complete the calculations by using the sequence Flash, Split, and React to solve the process. Any stream selected as a "forward or backward recycle" is termed a *tear stream*. Aspen Plus has a facility to enable the user to select tear streams and to choose the sequence of calculations. These options can be found under Convergence, Tear, Convergence, ConvOrder, and Convergence, Sequence. Whether Aspen Plus or the user selects the tear streams, it is necessary to initiate the calculations with a set of starting values. Aspen Plus provides zeros, which in many situations will lead to a converged solution. If it does not, the user must provide reasonable estimates by entering values on the stream input form. Prior to rerunning the model, the calculations should be reinitialized by selecting the Run menu and clicking on Reinitialize.

As an illustration of the method of direct iteration, consider solution of the equation

$$f(x) = x^2 - c \tag{4.1}$$

Add the quantity x to both sides of the equation to form the function $F(x)$,

$$F(x) = x^2 - c + x = f(x) + x \tag{4.2}$$

and look for a solution where

$$F(x) = x \tag{4.3}$$

which occurs when $f(x) = 0$. Repeated iterations yield the solution desired. The derivative $F'(x)$ can be evaluated, numerically, from two successive iterations. The process converges under the conditions

$$0 < F'(x) < 1 \tag{4.4}$$

$$-1 < F'(x) < 0 \tag{4.5}$$

Aspen Plus provides several other convergence options which may be selected by opening Data, Convergence, ConvOptions, Defaults, and the tab Default Methods. The default method (Wegstein, 1958) is usually suitable for most applications. Wegstein's method provides substantial robustness by making use of the first derivative and modifying the next trial value of x_2 as follows:

$$T = \frac{1}{1 - F'(x_1)} \tag{4.6}$$

$$x_2 = (1 - T)x_1 + TF(x_1) \tag{4.7}$$

The direct iteration and Wegstein's methods treat each of the state elements of a stream as if it were independent of the others, and hence for a single, n-component tear stream there are $n + 1$ equations in either method. Some of these ideas are illustrated in Figure 4.3, due to Rubin (1964), which has been modified minimally to facilitate the use of Aspen Plus's blocks. Using the traditional approach, the flowsheet appears to have five tear streams, 1, 3, 4, 6, and 7, using the calculation sequence Mix, B2, B3, B4, SP1, B5, and SP2. Selection of these streams as tears is illustrated as follows. The selection Convergence under the setup drop-down menu opens the object manager, and when New is selected, Figure 4.4 appears. It is then possible to enter the desired tear streams. Aspen Plus detects whether the sequence is feasible and issues error messages if it is not.

Through visual trial and error the reader should be able to discover the sequence B3, B4, SP1, B5, SP2, Mix, and B2 with tear streams 2 and 5, thus reducing the number of tear streams from five to two and potentially reducing the calculations considerably. As part of Rubin's paper he developed an algorithm that can produce the minimum number of tears required for an arbitrary flowsheet. Such an algorithm is a part of Aspen Plus, and when this problem was solved with Aspen Plus, the optimal sequence was identified. This problem is available for study at Chapter Four Examples/Rubin.bkp.

The well-known Newton method is an alternative method for solving process flow-sheets. Details of the method are given in Section 6.5. A significant difference between Newton and Wegstein is the formulation of equations that are to be solved. Each equation still has the form of equation (4.2), except that with Newton's method each of the $2n_c + 1$ equations is a function of the $2n_c + 1$ unknowns associated with that tear stream. For example, if there are n_t tear streams, $n_t(2n_c + 1)$ unknowns and equations are to be solved. The Newton method requires partial derivatives of each equation with respect to each variable for each iteration. To create these numerically, after Aspen Plus

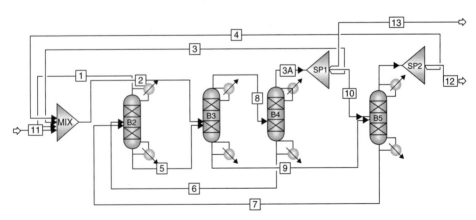

Figure 4.3   Rubin's flowsheet modified.

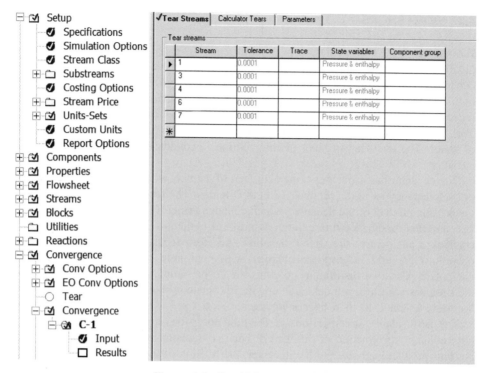

Figure 4.4 Specifying tear streams.

has an unconverged trial solution, each variable is in turn perturbed by a small amount; thus, $n_t(2n + 1)$ passes through the flowsheet are made. Convergence is quadratic.

The Broyden (1965) method is a variant of Newton's which estimates the values of the partial derivatives from the current values and may require fewer passes through the flowsheet. Convergence is linear.

4.2 HEURISTICS

1. Prior to attacking a material and energy balance problem, attempt a solution with simple blocks without energy balances.
2. Solve easy variations of a problem prior to attempting the actual problem (i.e., do not use tight specifications).
3. If errors are found in the solution and corrected, for the next run be sure to reinitialize the flowsheet; otherwise, Aspen Plus will continue the solution using the current, incorrect, values when it executes, which may prevent convergence.
4. For a large flowsheet or one with a complex configuration in which the tear streams are dependent on each other:
   a. If it fails to converge try solving with Aspen Plus's defaults.
   b. Break the flowsheet into smaller segments and converge them separately prior to integrating them into the complete configuration.

4.3 WORKSHOPS

Workshop 4.1   The separation of ethanol from water by heterogeneous azeotropic distillation is carried out by two distillation columns, each with its own overhead condensor but sharing a common decanter, as shown in Figure 4.5. The first column has two feeds, one of fresh material and one containing the water-rich phase recycled from the decanter. The bottoms product of this column is primarily water. The fresh ethanol-rich feed is composed of 1000 lb/hr of ethanol and 9000 lb/hr of water. The overhead product of this column is condensed and sent to the shared decanter.

The decanter has a fresh cyclohexane feed of 1 lb/hr, which represents the makeup of cyclohexane lost in the products of both columns. The feed to the second column is the organic product of the decanter, which contains primarily ethanol and cyclohexane. The bottoms product of this column is primarily ethanol. The overheads product is condensed and sent to the shared decanter. All fresh feeds are saturated liquids. All operations are at 1 atmosphere pressure with negligible pressure drops within the equipment. The two distillation columns are to be simulated with Sep2 blocks and the decanter with a Sep block. Solve using the material-balance-only option and use Wegstein's default method for convergence.

The performance specifications of the unit operations are given in Table 4.1. For both columns the fraction of the feed of each component leaving with the respective bottoms product is given. For the decanter, the fraction of the feed of each component leaving with the organic phase is given.

a. Why may one assume that the system can be represented by ideal thermodynamics for the vapor and liquid phases?

b. Has Aspen Plus chosen the best tear streams?

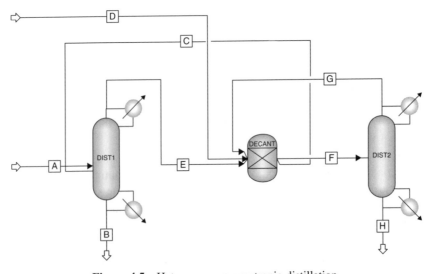

Figure 4.5   Heterogeneous azeotropic distillation.

TABLE 4.1 Process Performance Data

Component	Fraction of Feed to Bottoms		Fraction of Feed to Organic Phase: Decanter
	Column 1	Column 2	
Ethanol	0.01	0.97	0.98
Water	0.97	0.0001	0.01
Cyclohexane	0.09	0.0001	0.99

Workshop 4.2 Increase the number of Wegstein iterations to 100. Reinitialize the problem, and solve. What do you observe?

Workshop 4.3 Select C and F as the recycle streams. Provide initial values and reinitialize.

a. Solve using Wegstein's method. What do you observe?
b. Solve using Newton's method. What do you observe?

Workshop 4.4 Figure 4.6 shows a continuous process for the recovery of solvents from a mixture that includes a soluble solid. Upon concentration, the solid precipitates and can be removed by filtration. You are to perform preliminary calculations without energy balances using the process description given below.

The primary feed to the process, FD1, is composed of 100 lb/hr of ethanol (ETOH), 750 lb/hr of water (WATER), 100 lb/hr of toluene (TOL), and 50 lb/hr of soluble solids (SOLIDS) at 14.696 psi. FD1 is sent to a column (CRUDSTL) along with a second feed, FD2, composed of 250 lb/hr of saturated steam (H_2O) at 30 psig. The still produces two products: SOLVNTS and CRSOLIDS. 95% of the ethanol fed, 99% of the toluene fed, and 50% of the water fed end up in the stream SOLVNTS, and all of the solids end up in the stream CRSOLIDS. All of the solids in this stream precipitate out of solution.

The stream CRSOLIDS is the feed to a filter that produces two streams: AQSLVNT and SOLIDS. All of the solids fed, 1.0% of the ethanol fed, 1.0% of the toluene fed, and 0.5% of the water fed end up in the stream SOLIDS.

The stream SOLVNTS is fed to a column (TOLSTILL) whose role is to separate ethanol and toluene. The overheads (CRUDETOH) from this still are a stream that is primarily ethanol and water which is purified in the azeotropic distillation columns ESTILLHI and ESTILLLO. The bottoms (CRUDTOL) are sent to a decanter (DECANT). The performance characteristics of TOLSTILL are as follows: 25% of the ethanol fed, 45% of the water fed, and 99% of the toluene fed end up in the stream CRUDTOL.

The decanter (DECANT) produces a relatively pure toluene product (TOL) and an aqueous product (AQTOL), which is recycled to the column TOLSTILL. The performance characteristics of the decanter are as follows: 50% of the ethanol fed, 99.9% of the water fed, and 0.1% of the toluene fed end up in the stream AQTOL.

The stream CRUDETOH feeds the column ESTILLHI. The column produces two products: ETAZH, whose composition is to be 88 mol % ethanol, and ETAQ, whose

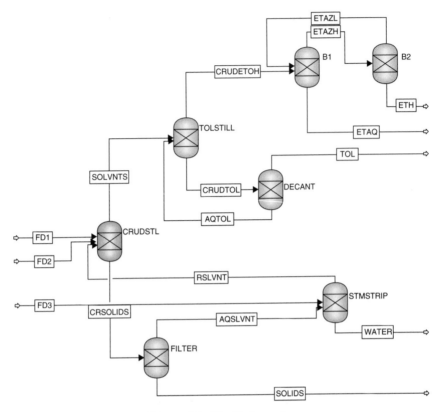

Figure 4.6 Distillation train.

composition is to be 99% water. All of the toluene fed ends up in the stream ETAQ. The stream ETAZH feeds the column ESTILLLO. This column produces the relatively pure product ETH, which is to contain 99 mol % ethanol, and the product ETAZL, which is to contain 82 mol % ethanol. The stream ETAZL is recycled to column ESTILLHI.

A steam stripper (STMSTRIP) is provided to remove solvents from the aqueous products of the process, ETAQ and AQSLVNT. The stream FD3 consists of 50 lb/hr of 30-psig steam that is used as the energy source for the stripping the solvents. The stripper produces the stream RSLVNT, which contains the stripped solvents and is recycled to the column CRUDSTL. The stream WATER is discharged to the environment. The performance characteristics of the stripper are as follows: 0.5% of the ethanol fed, 90% of the water fed, and 0.01% of the toluene fed end up in the stream WATER.

a. There is no component SOLIDS. Devise a method that will permit this component to be used in the simulation.

b. To calculate the material balance for the complete process flow sheet, it may be advisable to break up the process into manageable segments that are to be combined when the complete material balance is solved. Select the segments and solve each using an estimate for the required feeds when necessary.

c. Solve the complete process. Take care to place starting values for the tear streams estimated from the segments.

Workshop 4.5 Search Aspen Plus Help for adding a stream table to your results. Replace all simple boxes with suitable icons. Place a stream table on your flowsheet graphic. Place suitable text labels on your flowsheet.

Workshop Notes

Workshop 4.1

a. Since this is a material-balance-only approach, no rigorous blocks are used, and all internal calculations are by simple blocks, the type of thermodynamics chosen is arbitrary since it is not used in the simulation. If this problem were solved with rigorous blocks such as Radfrac and Decant, rigorous thermodynamics would be required.
b. The Aspen Plus execution does not converge. You may try the following:
 • Increase the number of iterations. Check the overall material balance.
 • Check the performance of each block.
 • Check that no unrealistic specifications were used.
 • Change the tear streams.
 • Change the convergence method used.

Workshop 4.2 Four-2.bkp uses 100 Wegstein iterations and it, too, fails to converge. This looks like a situation where Wegstein may not work or the tear streams Aspen chose were not suitable. Examining this process from a chemical engineering point of view, the composition of the streams that are easiest to estimate are those leaving the decanter: streams C and F.

Workshop 4.3 Streams C and F may be initialized with unconverged results from Four-2.bkp. As an alternative, the compositions of streams C and F can be estimated from a phase diagram of the ethanol–water–cyclohexane system.

a. Four-3.bkp using Wegstein's method again fails to converge. This problem is not suitable for Wegstein convergence, even with 100 iterations.
b. Four-3n.bkp using Newton's method converges easily with only a few iterations.

Workshop 4.4

a. Since Aspen Plus deals mostly with solids in a special way (an advanced application) and the states considered are either vapor or liquid, there is no specific provision for dealing with solids dissolved or suspended in a solution. Furthermore, for a situation in which there are no specific database components for unidentified by-products of reactions, one must assure that these are accounted for on a mass basis, and that their behavior in rigorous blocks is accounted for in some way. For example, it is safe to assume that the unknown vapor pressure of a solid by-product is much lower than that of the other components; therefore, none of that species will appear in a vapor phase. A simple method to accomplish this is to select a very heavy component, such as eicosane, $C_{20}H_{42}$, from the database and change its name, for example, to HEAVY, and ignore the properties that are calculated when that species is part of a mixture.

b. Select the sequences
   - CRUDSTL, FILTER, STMSTRP
   - TOLSTILL, DECANT
   - B1, B2

The complete solution may be found at Four-4.bkp. Note that if this were to be solved rigorously, many problems would be raised by the complex thermodynamics required. This problem is addressed rigorously in Chapter Fourteen.

Workshop 4.5 A solution to the flowsheet is required. Under the View menu, the PFD mode must be checked.

REFERENCES

Aspen Plus version 7.0, Help: Stream Table.
Broyden, C. G., *Math. Comput*. 19(92), 577–593 (1965).
Rubin, I. D., *Chem. Eng. Symp. Ser. Appl. Math. Chem. Eng*., 55(37), 54–62 (1964).
Wegstein, J. H., Accelerating convergence of iterative processes, *Commun. ACM*, 1, 9 (1958).

FLOWSHEETING AND MODEL ANALYSIS TOOLS

Two of the most useful tools in Aspen Plus are the Sensitivity study and the Design specification, both located on the Data Browser. These tools work hand-in-hand and are used when a design specification is used to specify a block output variable. For example, the file Chapter Five Examples/mixer5.bkp presents a mixer block with three feeds. When all input streams are specified properly, the block will calculate the composition and state of the output stream. Figures 5.1 and 5.2 show the graphic setup and the solution. Note that the product, stream 4, is a straightforward summation on a componental basis, of all the feeds, excluding an energy balance. Suppose, however, that in the example one requires that the composition of a specific component, say ethanol, is to be 0.5 mole fraction. It will be necessary to increase the flow of stream 2. This is easy to do with a hand calculation, but if this block were in the middle of a process flow diagram, it might not be a trivial task. The Sensitivity study permits the specification of a trial range of flow rates for stream 2, which may be plotted against the resulting ethanol composition. Once an approximate solution is obtained, by plotting the results and checking the location of the target composition, the Design specification can be employed to provide a means for Aspen Plus to solve for the exact value of the flow rate of stream 2, which satisfies the specification desired. In setting up both the Sensitivity study and the Design specification, it may be necessary to employ the Fortran language capability embedded in several blocks and available on a stand-alone basis in the Calculator block.

5.1 INTRODUCTION TO FORTRAN IN ASPEN PLUS

Fortran (Formula Translation) is a very old but rich language which has traditionally been used for scientific applications. Details of the language are very complex and can

Teach Yourself the Basics of Aspen Plus™ By Ralph Schefflan
Copyright © 2011 John Wiley & Sons, Inc.

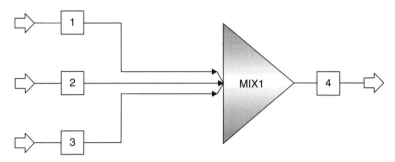

Figure 5.1   Mixer5 example.

	1	2	3	4
Substream: MIXED				
Mole Flow lbmol/hr				
METHANOL	100.0000	0.0	0.0	100.0000
ETHANOL	0.0	100.0000	0.0	100.0000
WATER	0.0	0.0	100.0000	100.0000
Mole Frac				
METHANOL	1.000000	0.0	0.0	.3333333
ETHANOL	0.0	1.000000	0.0	.3333333
WATER	0.0	0.0	1.000000	.3333333
Total Flow lbmol/hr	100.0000	100.0000	100.0000	300.0000
Total Flow lb/hr	3204.216	4606.904	1801.528	9612.648
Total Flow				
Temperature F	70.00000	70.00000	70.00000	
Pressure psia	14.70000	14.70000	14.70000	14.70000

Figure 5.2   Mixer5 solution.

be found in many books. Aspen Plus employs both interpreted Fortran (line-by-line execution) for the blocks in this chapter and compiled Fortran (dynamic link library creation) for the creation of user models. An overview of the language criteria can be found in Aspen Plus Help and by searching About the Interpreter. Aspen Plus user model documentation recommends use of the Intel Fortran compiler.

5.2   BASIC INTERPRETED FORTRAN CAPABILITIES

For the purposes of this chapter's functions, what follows is the primary function-ality that is available when using interpreted Fortran. There is no intent here to present a thorough documentation of the Fortran programming language, merely what is sufficient to be able to work with the subject matter of this chapter and some material in later chapters.

5.2.1 Primary Fortran Operators

The primary Fortran operators are given in Table 5.1. Prior to execution of any statement, all variables appearing on the right-hand side of an equal sign must be defined: that is, contain either a numerical value or a value calculated with previous legal statements.

A list of all the functions available is given in the Aspen Plus documentation About the Interpreter.

5.2.2 Precedence of Calculations

The following rules apply when a Fortran statement is evaluated:

- Fortran statements may be organized in groups of operations within parentheses.
- The contents of each group is evaluated prior to evaluation of the complete statement.
- Function calls precede group evaluation.
- Exponentiation precedes group evaluation.
- If part of a group, divisors are evaluated prior to numerators.

For example, the precedence of calculations for the Fortran statement

$$a = \frac{(f * dlog(b) + c)}{(a + b) + (c - d * e ** f)} \tag{5.1}$$

is as follows:

TABLE 5.1   Fortran Operators

Operator	Definition	Example	Result
=	replacement	A = B	B
+	addition	A = A + A	2A
−	subtraction	A = A − A	0
/	division	A = A/A	1
*	multiplication	A = A*A	A^2
**	exponentiation	A = A**B	A^B
.LT.	logical less than		
.GT.	logical greater than		
.EQ.	logical equal		
..AND.	logical and		
	function call	A = Dlog(B)	Ln(B)
	subroutine call	CALL XYZ	XYZ executes

1. $e^{**}f$
2. $d^*e^{**}f$
3. $(c-d^*e^{**}f)$
4. $(a+b)$
5. $(a+b)+(c-d^*e^{**}f)$
6. $dlog(b)$
7. $f^*\ dlog(b)$
8. $(f^*dlog(b)+c)$
9. $(f^*dlog(b)+c)/(a+b)+(c-d^*e^{**}f)$

5.2.3 Statement Format

Intel Fortran, as implemented in Aspen Plus's interpreter, is not quite standard (there are differences of opinion as to what constitutes standard Fortran). The main formatting rules, taken from Aspen Plus's documentation, are given below.

- Column 1 can contain a C. All other entries on the same line are comments.
- Column 2 is left blank.
- Columns 3 to 5 are reserved for statement labels. Under certain logical conditions, program execution can be transferred to a labeled statement.
- Executable statements begin at column 7 and beyond.
- No variables may begin with the characters IZ or ZZ.
- Integer variables begin with the letters I through N.
- Real variables begin with the letters A through H or O through Z.
- Variables names are limited in length to seven characters.
- Lowercase characters are permitted.

5.2.4 Program Logic Control

Logical statements may take a variety of forms. An example of one of the IF variations, the most common, is as follows:

```
IF ((expression a). EQ. (expression b)) (expression c)
        expression d
```

Expression c will execute if the logical grouping is true; otherwise, expression d executes. In all statements of this type, the number of left parentheses must equal the number of right parentheses. The logical operators are all interchangeable and can be nested within each expression to meet program requirements.

Another common logic control statement is the direct Go To:

```
Go To statement label
```

All statements following the Go To are skipped until the labeled statement is found; then execution proceeds normally.

5.3 SENSITIVITY FUNCTION

The features of the sensitivity function will be displayed by the continuation of example mixer5.bkp through example mixer5s.bkp. The Sensitivity function is located under the setup menu Model Analysis Tools and when selected initiates the Sensitivity Data Browser. When New is selected, a Sensitivity ID such as S-1 is assigned. This initiates a display that facilitates the association of a Fortran-accessible variable with a variable within the process. It is imperative that this process be started by selecting the New button, which generates the displays shown in Figure 5.3. The variable xetoh, which is user selected, represents the mole fraction of ethanol in output stream 4 of the example. Figure 5.4 shows the assignment of the process variable to the Fortran variable. For this example, a Fortran expression will be created to calculate the fraction of alcohols in stream 4. For this purpose it will be necessary to create three additional Fortran-accessible variables (fmoh, fetoh, and f4), the flow rates of methanol and ethanol in stream 4, and the total flow of stream 4.

Selection of the tab Vary in the Sensitivity input Data Browser brings up Figure 5.5, which defines the range of trial values (the independent variable) that stream 2 should use in seeking for a value that would produce an approximate solution to the desired target of ethanol mole fraction in stream 4 of 0.5.

Selection of the tab Tabulate permits the assignment of any defined Fortran variables or legitimate Fortran expression to the column of a table on which the results of the Sensitivity study will be presented. Note that column 3 contains an expression for the calculation of the fraction of alcohols leaving the mixer. Tabular and plotted results are displayed in Figure 5.6. One can see that the flow rate required to produce a 0.5 mole fraction of ethanol is bracketed between a stream 2 flow of 100 and 300 mol/hr, and from the plot the solution is approximately 225 mol/hr. Note that the mole

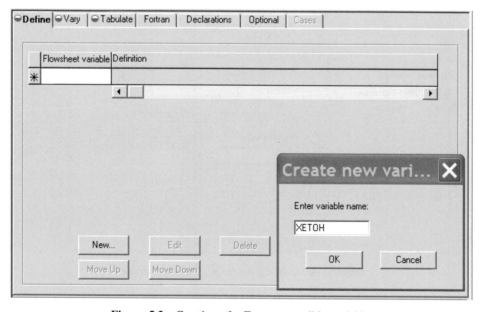

Figure 5.3   Creation of a Fortran-accesible variable.

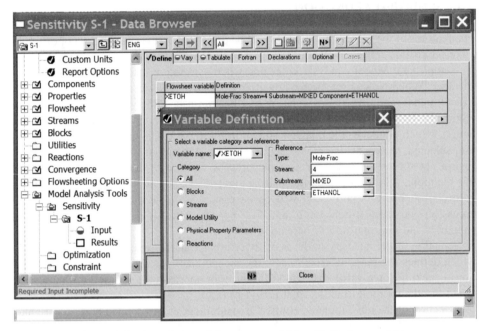

Figure 5.4   Association of flowsheet variable with Fortran variable.

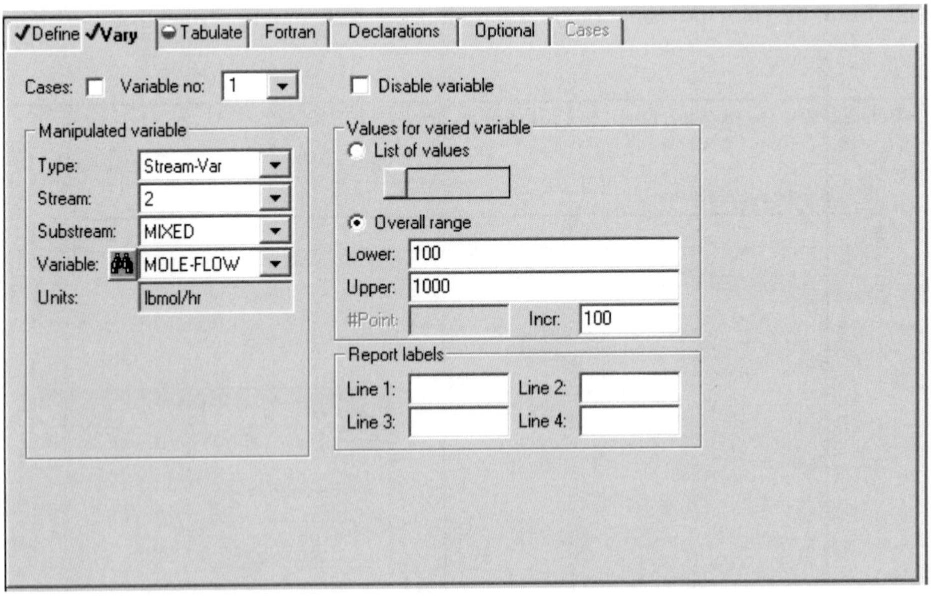

Figure 5.5   Defining the range of a trial variable.

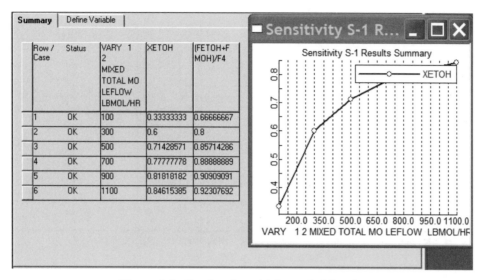

Row / Case	Status	VARY 1 2 MIXED TOTAL MO LEFLOW LBMOL/HR	XETOH	(FETOH+F MOH)/F4
1	OK	100	0.33333333	0.66666667
2	OK	300	0.6	0.8
3	OK	500	0.71428571	0.85714286
4	OK	700	0.77777778	0.88888889
5	OK	900	0.81818182	0.90909091
6	OK	1100	0.84615385	0.92307692

Figure 5.6   Results of sensitivity study Mixer5s.

fraction of alcohols in stream 4 is tabulated. This example may be found at Examples/
mixer5s.

5.4 DESIGN SPECIFICATION

The features of the design function will be displayed by the modification of the example
mixer5s.bkp and is given at mixer5d.bkp. The Design-Spec is located under the setup
menu Flowsheeting Options, and when selected initiates the Design-Spec Data Browser.
Selection of the New button assigns an identifier to a design spec such as DS-1. The
key to implementing the design specification is to manipulate the feed, stream 2, so that
it can take on the required flow rate to achieve the required stream 4 specification. One
possibility is shown in Figure 5.7. The inclusion of stream 5, which has a larger flow
than what is required to meet the stream 4 specification (the flow rate is determined
from the sensitivity study), is fed to a multiply block. The design spec function will
manipulate the multiplier value, which will place the appropriate value into stream 2 to
satisfy the design specification. The implementation requires the definition of a Fortran
variable, in this case, XETOH, just like the setup in the Sensitivity study; however,
there are two other tabs in the Design-Spec Data Browser, the target value and an
acceptable tolerance, which must be selected and implemented. Figure 5.8 shows the
setup of the function to be solved:

$$f(factor) = xetoh - 0.5 \tag{5.2}$$

where *factor* is the multiplying factor of the block M1. The function is solved by
Aspen Plus's built-in secant method equation solver within a user-specified tolerance,
in this case 0.001. Results are shown in Figure 5.9.

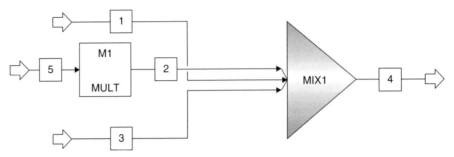

Figure 5.7 Flowsheet for Mixer5c.

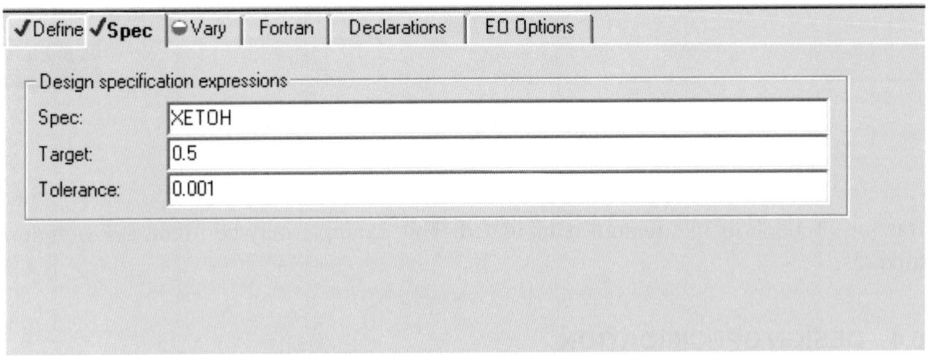

| ✓Define | ✓**Spec** | ⊖Vary | Fortran | Declarations | EO Options |

Design specification expressions

Spec:	XETOH
Target:	0.5
Tolerance:	0.001

Figure 5.8 Mixer5d design specification.

| Display: All streams ▼ | Format: FULL ▼ | Stream Table |

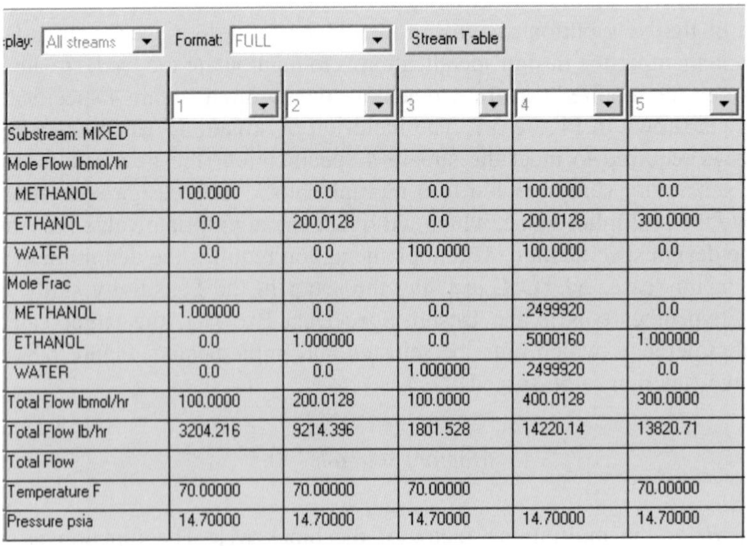

	1 ▼	2 ▼	3 ▼	4 ▼	5 ▼
Substream: MIXED					
Mole Flow lbmol/hr					
METHANOL	100.0000	0.0	0.0	100.0000	0.0
ETHANOL	0.0	200.0128	0.0	200.0128	300.0000
WATER	0.0	0.0	100.0000	100.0000	0.0
Mole Frac					
METHANOL	1.000000	0.0	0.0	.2499920	0.0
ETHANOL	0.0	1.000000	0.0	.5000160	1.000000
WATER	0.0	0.0	1.000000	.2499920	0.0
Total Flow lbmol/hr	100.0000	200.0128	100.0000	400.0128	300.0000
Total Flow lb/hr	3204.216	9214.396	1801.528	14220.14	13820.71
Total Flow					
Temperature F	70.00000	70.00000	70.00000		70.00000
Pressure psia	14.70000	14.70000	14.70000	14.70000	14.70000

Figure 5.9 Mixer5d design specification satisfied.

5.5   CALCULATOR FUNCTION

The calculator may be found under the Setup menu Flowsheeting options. Like the Design and Sensitivity functions, the user associates flowsheet variables with user-defined Fortran variables. The basic calculator display is shown in Figure 5.10. The association of a Fortran variable occurs after the New button is clicked, but requires that the line upon which the new variable is entered be selected and the Edit button clicked. The entries for the creation of the variable FLO2 are shown in Figure 5.11. Two very important selections are Import, which copies current values of variables from the flowsheet into the Fortran variables, and Export, which copies values of the Fortran variables into current variables of the flowsheet. An example of Fortran coding that accomplishes the same result as in the example Mix5d.bkp is given in Figure 5.12, which uses the definitions in Figure 5.10. Figure 5.13 illustrates the specifications, which permit placement of the calculation virtually anywhere in the sequence of flowsheet calculations.

The results of the calculations are shown in Figure 5.14. The results were accomplished with the first Fortran statement, a material balance to calculate the amount of ethanol that must be added to stream 4 to make the 0.5 mole fraction specification. The second statement exports the sum of the original value of ethanol in stream 4 and the results of the first statement. When these operations are completed, Aspen Plus issues a series of errors and warnings which are documented in the control panel as shown in Figure 5.15. This example may be found at Examples/mixer5c. Attention should be paid to the information in the control panel because it will caution the user to any irregularity. As an example, note the warning that the mixer block is not in material balance. This occurred because the mixer was executed prior to the calculation function.

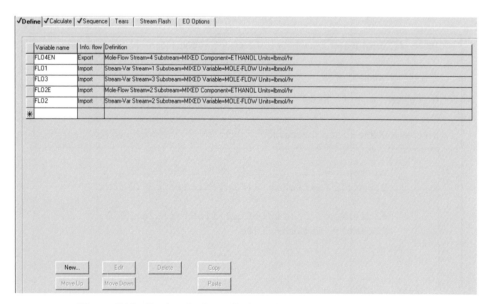

Figure 5.10   Basic calculator display with Fortran variables defined.

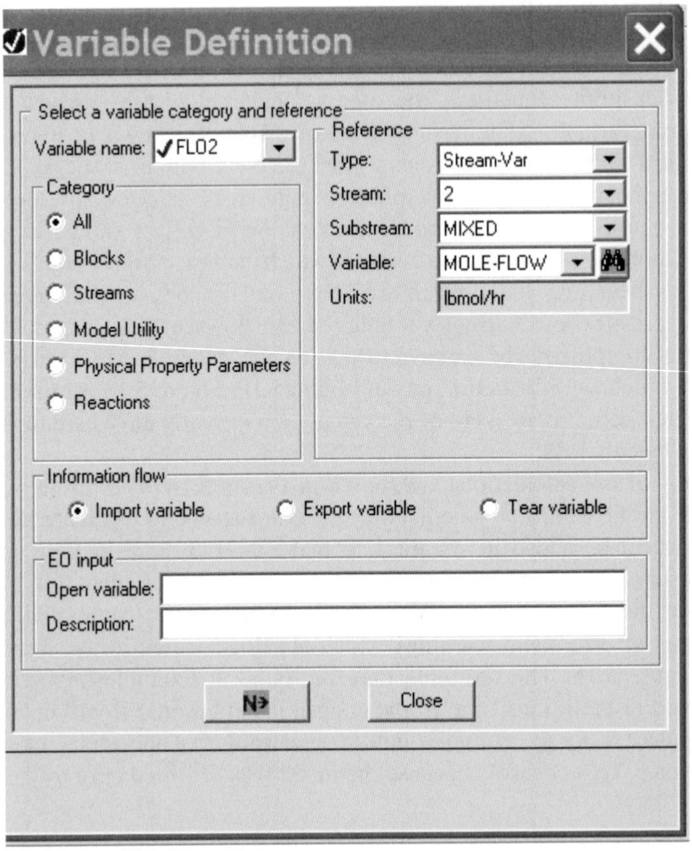

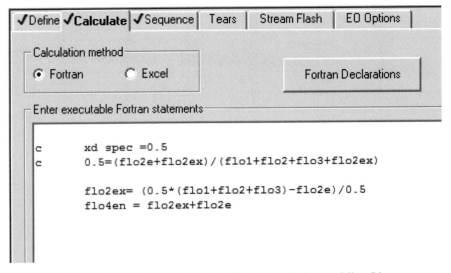

Figure 5.11 Association of flowsheet variable with Fortran variable.

```
c       xd spec =0.5
c       0.5=(flo2e+flo2ex)/(flo1+flo2+flo3+flo2ex)

        flo2ex= (0.5*(flo1+flo2+flo3)-flo2e)/0.5
        flo4en = flo2ex+flo2e
```

Figure 5.12 Fortran coding for Mixer5c equivalent to Mixer5d.

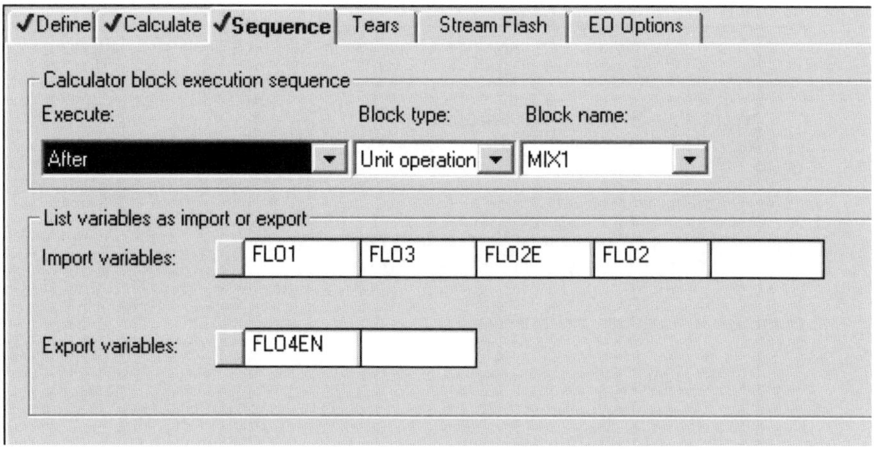

Figure 5.13 Sequence control of Fortran code.

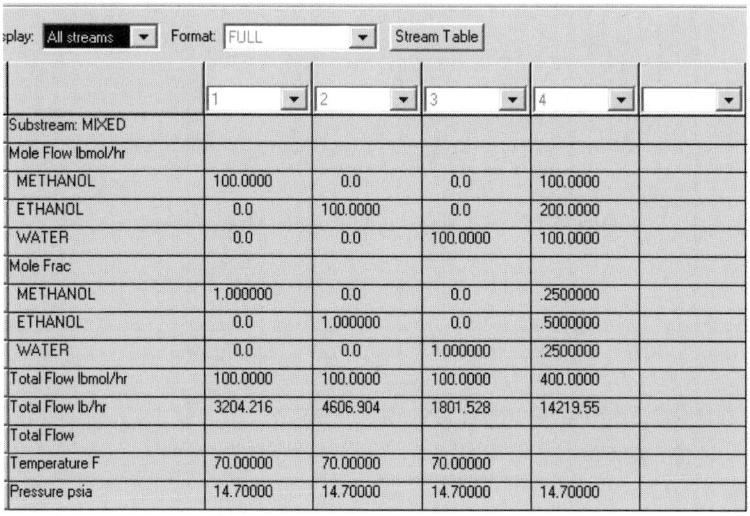

	1	2	3	4	
Substream: MIXED					
Mole Flow lbmol/hr					
METHANOL	100.0000	0.0	0.0	100.0000	
ETHANOL	0.0	100.0000	0.0	200.0000	
WATER	0.0	0.0	100.0000	100.0000	
Mole Frac					
METHANOL	1.000000	0.0	0.0	.2500000	
ETHANOL	0.0	1.000000	0.0	.5000000	
WATER	0.0	0.0	1.000000	.2500000	
Total Flow lbmol/hr	100.0000	100.0000	100.0000	400.0000	
Total Flow lb/hr	3204.216	4606.904	1801.528	14219.55	
Total Flow					
Temperature F	70.00000	70.00000	70.00000		
Pressure psia	14.70000	14.70000	14.70000	14.70000	

Figure 5.14 Results of Mixer5c.

5.6 TRANSFER FUNCTION

The error "mixer in the example mixer5c.bkp is not in material balance" occurs because stream 2 has not been updated with the contents of stream 4, as shown in Figure 5.14. Aspen Plus provides the Transfer function, which enables the copying of data from stream to stream. When the Transfer function is invoked, an ID is assigned and the display shown in Figure 5.16 is displayed. Selection of tabs enables the selection of streams to copy from and to as well as which elements of those streams to select. The tab sequence enables the user to choose the order within the flowsheet calculation to run the function. Figure 5.17 shows the results of the transfer.

```
->Processing input specifications ...

  *   WARNING DURING FLOWSHEET ANALYSIS
      FORTRAN WRITE-VAR FLO4EN IN FORTRAN BLOCK C-1
      IS WRITING TO AN OUTLET STREAM VARIABLE.

  Flowsheet Analysis :
  *   WARNING DURING FLOWSHEET ANALYSIS
      Sequence specification (C-1 EXECUTE AFTER MIX1)
      may be inconsistent with variable accessing.


COMPUTATION ORDER FOR THE FLOWSHEET:
MIX1 C-1

->Calculations begin ...


  Block: MIX1     Model: MIXER

  Calculator Block C-1

 **  ERROR
      BLOCK MIX1 IS NOT IN MASS BALANCE:
      MASS INLET FLOW = 0.12111733E+01, MASS OUTLET FLOW = 0.17916334E+01
      RELATIVE DIFFERENCE = 0.47925442E+00
      A STREAM FLOW MAY HAVE BEEN CHANGED BY A FORTRAN, TRANSFER,
      OR BALANCE BLOCK AFTER THE BLOCK HAD BEEN EXECUTED.


->Simulation calculations completed ...
```

Figure 5.15 Warnings and errors in Mixer5c results.

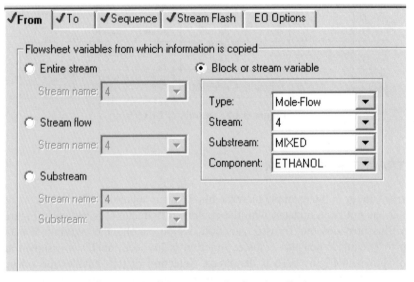

Figure 5.16 Primary transfer function display.

	1	2	3	4
Substream: MIXED				
Mole Flow lbmol/hr				
METHANOL	100.0000	0.0	0.0	100.0000
ETHANOL	0.0	200.0000	0.0	200.0000
WATER	0.0	0.0	100.0000	100.0000
Mole Frac				
METHANOL	1.000000	0.0	0.0	.2500000
ETHANOL	0.0	1.000000	0.0	.5000000
WATER	0.0	0.0	1.000000	.2500000
Total Flow lbmol/hr	100.0000	200.0000	100.0000	400.0000
Total Flow lb/hr	3204.216	9213.808	1801.528	14219.55
Total Flow				
Temperature F	70.00000	70.00000	70.00000	
Pressure psia	14.70000	14.70000	14.70000	14.70000

Figure 5.17   Results of Mixer5t.

5.7 WORKSHOPS

Workshop 5.1   Copy the file four3n.bkp, which will be the basis for using Aspen Plus's Sensitivity Studies capability to determine the effect of the decanter's performance with regard to the fraction of cyclohexane leaving the decanter in its aqueous product, stream C. For the base case this specification is 0.02. The Sensitivity Studies should vary this value between 0.01 and 0.10 in increments of 0.01. Select as tear streams C and F. Make a reasonable estimate of the componential flows and solve the flowsheet using the Newton convergence option. This becomes the base case solution. This file is to be modified using the Sensitivity function. Select the following as the dependent variables:

1. The total flow of stream G, the cyclohexane-rich distillate recycled to the decanter. Call it FLOWG.
2. The flow of cyclohexane in stream G. Call it FLOWGC.

Plot the results. If the pipeline in which G is to flow is limited to 300 lb/hr, what is the fraction of the cyclohexane entering the decanter that leaves in stream C?

Workshop 5.2   Using Workshop 5.1 as a basis for solving Workshop 5.1 by means of the Design Specification capability of Aspen Plus, delete the Sensitivity function and replace it with Design-Spec. Vary the decanter performance with regard to cyclohexane in stream C and use it as brackets for the iterations 0.01 and 0.10. Set up the design specification such that the total flow rate of stream G is 300 lb/hr with a tolerance of 0.1 lb/hr. What is the fraction of the cyclohexane entering the decanter that leaves in stream C?

Workshop 5.3 Using Workshop 5.1 as a basis, create a Sensitivity study to determine the effect of feed composition on the concentration of ethanol in stream B and water in stream H. To do this it is necessary to change the quantity of ethanol in stream A while reducing the amount of water in stream A by the same amount for each case. Aspen Plus's Sensitivity function will permit variation in the ethanol quantity for each case, but it will be necessary to write Fortran that executes prior to each time the Distillation I block executes. This capability will be found under "Flowsheeting Options—Calculator." Use the following values of ethanol in A, in units of lb/hr, for your Sensitivity studies: 1000, 1100, 1500, 2000, and 5000.

Workshop 5.4 Part of a plant-wide environmental control facility (ECF) is a system for diverting aqueous waste fluids from a discharge point such as a river in the event that the concentration of volatile organic compounds (VOCs) exceeds a certain level. This logic is to be modeled by employing a splitter with two product streams. If the composition of the VOCs exceed 20 parts per billion, the stream is to be diverted to the second outlet of the splitter. Otherwise, the stream is to be diverted to the first outlet. Execute a calculation block in front of the splitter to program the logic in Fortran. Test with the two potential feeds listed in Table 5.2.

TABLE 5.2 ECF Feeds

Component	VOC	Feed 1 lb/hr	Feed 2 lb/hr
Methanol	Yes	1	10
Acetone	Yes	2	20
Toluene	Yes	3	30
Methylene chloride	Yes	4	40
Water	No	1.0×10^9	1.0×10^9

Workshop Notes

Workshop 5.3 The Fortran statement required may be found in the file Workshops/Five-3.txt.

Workshop 5.4 The Fortran statements required may be found in the file Workshops/Five-4a.txt.

REFERENCES

Aspen Plus version 7.0, About the Interpreter documentation.
Intel Fortran Programmer's Reference, http://docs.nscl.msu.edu/ifc/intelfor_prglangref.pd, 1996–2003.

CHAPTER SIX

THE DATA REGRESSION SYSTEM

Aspen Plus's regression system can be used to fit various types of pure component physical property data such as vapor pressure data, or a parameter for a new component in the Unifac database, but its primary use is to customize thermodynamic models dealing with vapor–liquid equilibrium (VLE) and liquid–liquid equilibrium (LLE). Most of the equations of state and activity coefficient models commonly used are described in detail by Poling et al. (2000), as are listings of textbooks and data sources relating to fluid-phase equilibrium. Additional descriptions of models and variants may be found in Aspen Plus's documentation.

When an ideal system is modeled, the equations and/or correlations that are used to calculate the physical and thermodynamic properties of both phases are well defined and do not require augmentation with experimental data; however, the opposite is true when a real system is considered. In this regard there are two possibilities:

1. When both phases of a system can be represented by an equation of state, the fundamental equation for vapor–liquid equilibrium for component i is given by

$$y_i \phi_i^V = x_i \phi_i^L \tag{6.1}$$

   where, y and x are mole fractions of the vapor and liquid phases, respectively, and ϕ is a fugacity coefficient which depends on temperature, pressure, and composition and is calculated from an equation of state.

2. When the vapor phase can be represented by an equation of state and the liquid phase by an activity coefficient equation, the fundamental equation for vapor equilibrium is

$$y_i \phi_i P = \gamma_i x_i f_i^0 \tag{6.2}$$

where P is pressure, ϕ is calculated from an equation of state, γ is an activity coefficient calculated from an activity coefficient equation, and f_i^0 is the standard state fugacity, which is given by

$$f_i^0 = \phi_i^{\text{sat}} P_i^{\text{sat}} \exp \left(\int_{P^{\text{sat}}}^{P} \frac{v^L}{RT} dP \right) \tag{6.3}$$

The exponential term above, known as the *Poynting correction*, is used for very light components and is usually negligible. In most cases the vapor pressure, P^{sat}, is used as the componential liquid-phase fugacity.

The fundamental equation for liquid–liquid equilibrium equates two instances of the right-hand side of equation (6.2) written once for each liquid phase.

Aspen Plus supplies property methods identified by a specific vapor-phase fugacity equation and a specific activity coefficient equation which defines the use of equation (6.1) or (6.2). For example, the method NRTL-2, which represents the nonrandom two-liquid activity coefficient equation, also known as the Renon equation, is used in conjunction with the ideal gas equation and employs equation (6.2) in calculations. The "2" identifies the data set from which the NRTL parameters were derived. Aspen Plus has the possibility of two sets of activity coefficient parameters for most activity coefficient equations with an indicated range of applicability (i.e., temperature and pressure). Henry's law is used in some property methods for light components with Henry's law parameters provided in the Aspen Plus database. Aspen Plus supplies more than 50 property methods.

6.1 PARAMETERS OF EQUATIONS OF STATE

The most commonly used nonideal equations of state available in Aspen Plus all have the same basic structure: third order. The general formulation from which all modern forms can be derived is the five-parameter form

$$P = \frac{RT}{V - b} - \frac{\theta(V - \eta)}{(V - b)(V^2 + \delta V + \varepsilon)} \tag{6.4}$$

The parameters a and b are functions of the critical properties and the acentric factor ω. The symbols θ, η, δ, and ε have values that depend on the equation of state (chosen details may be found in Poling, 2000). For example, the values for pure components of the Peng–Robinson equation are

$$\theta = a\alpha(T_r)^* \tag{6.5}$$

$$\delta = 2b \tag{6.6}$$

$$\varepsilon = -b^2 \tag{6.7}$$

$$\alpha(T_r)^* = [1 + (0.37464 + 1.5422\,\omega - 0.2699\,\omega^2)(1 - T_r^{0.5})]^2 \tag{6.8}$$

For mixtures, the most commonly used equations of state are ideal, Redlich–Kwong, Soave–Redlich–Kwong with variants, and Peng–Robinson with variants. Most use the mixing rules given by

$$a = \sum_i \sum_j x_i x_j \sqrt{a_i a_j}(1 - k_{i,j}) \qquad (6.9)$$

$$b = \sum_i x_i b_i \qquad (6.10)$$

Modifications of the mixing rules for some variants may be found in Aspen Plus documentation. The introduction of the k_{ij} parameter, one for each binary pair in a mixture, improves the ability of the equations of state to fit experimental data. These values, which are available from a variety of sources (e.g., Knapp et al., 1982), are included in the Aspen Plus database and may be used within the temperature and pressure ranges given in Aspen Plus's documentation. There is rarely a need to fit the k_{ij} parameter. As an example, Figure 6.1 shows the k_{ij} parameters for the six binary pairs in the system nitrogen–methane–propane–n-pentane. Note that the upper and lower bounds of the working temperature ranges are given.

If the need arises to correlate a binary system not in the database, a set of *PVT* data with system vapor and liquid compositions is required. Aspen Plus implements the k_{ij} parameter with two terms:

$$k_{ij} = a_{ij} + b_{ij}T \qquad (6.11)$$

with the introduction of the $b_{i,j}$ term for temperature dependency. Some variants have additional terms which may be selected during the setup of the regression.

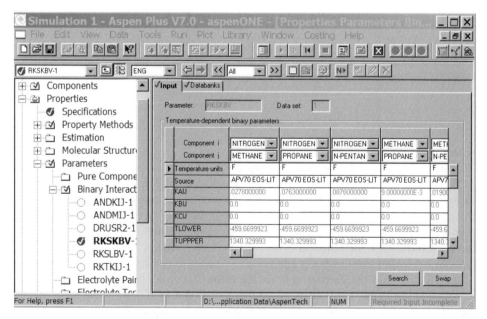

Figure 6.1   Soave–Redlich–Kwong Parameters.

6.2 PARAMETERS OF ACTIVITY COEFFICIENT EQUATIONS

Activity coefficient equations are derived from the general formulation of excess free energy,

$$RT \ln \gamma_i = \left(\frac{\partial G^E}{\partial n_i} \right)_{T,P,n_{j \neq i}} \tag{6.12}$$

where n_i is the number of moles of component i, and $n_{j \neq i}$ indicates any component except i. Applying equation (6.12) to a model of g^E, the total excess free energy per mole of mixture produces the activity coefficient equation. This is illustrated by the application of equation (6.12) to

$$\frac{g^E}{RT} = -x_1 \ln(x_1 + \Lambda_{12}x) - x_2 \ln(x_2 + \Lambda_{21}x_1) \tag{6.13}$$

the two-component formulation of g^E for the Wilson equation, and results in

$$RT \ln \gamma_1 = -x_1 \ln(x_1 + \Lambda_{12}x) - x_2 \left(\frac{\Lambda_{12}}{x_x + \Lambda_{12}x_2} - \frac{\Lambda_{21}}{\Lambda_{21}x_1 + x_2} \right) \tag{6.14}$$

where the activity coefficient for component 1 is γ_1, and Λ_{12} and Λ_{21} are parameters which are determined by regressing experimental data. When used in multicomponent calculations, binary parameters representing all permutations of the components are used in the multicomponent form of the Wilson equation:

$$\ln \gamma_i = -\ln \left(\sum_j^N x_j \Lambda_{ij} \right) + 1 - \sum_k^N \frac{x_k \Lambda_{ki}}{\sum_j^N x_j \Lambda_{kj}} \tag{6.15}$$

The Aspen Plus implementation of the Wilson parameter Λ_{ij} includes the user's choice of terms for temperature dependency and is shown in

$$\ln A_{ij} = a_{ij} + \frac{b_{ij}}{T} + c_{ij} \ln T + d_{ij}T + \frac{e_{ij}}{T^2} \tag{6.16}$$

where in Aspen Plus's terminology, A replaces Λ.

When regressing data the user may select which parameters of equation (6.16) are to be regressed from the Properties/Regression/R-1/Input form parameter tab, shown in Figure 6.2. On each column of the central part of the display reading downward, the parameter is identified as binary, the name of the activity model used, the element number [which corresponds to the coefficients in equation (6.16), e.g., a_{ij} is 1], and the components are identified in order; for example, ethanol is 1 and water is 2, so column 1 refers to parameter a_{12} and column 2 refers to parameter a_{21}. The remainder of the input permits the user to omit, fix the value of, regress the parameter and assign an initial value, define the bounds of the regression, and set a scale factor such that all parameters have comparably scaled contributions to the regression. The last entry on the column is used for regression of equations of state.

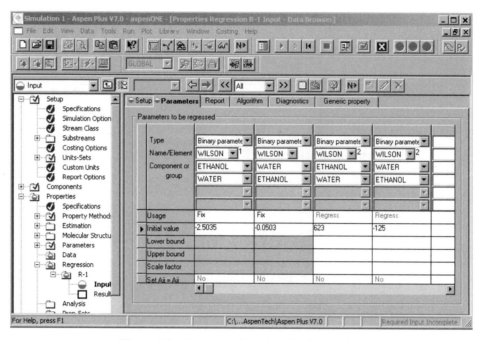

Figure 6.2   Parameter Selections for Regression.

The most commonly employed activity coefficient equations are the van Laar, Wilson, NRTL, Uniquac, and their variants. The input for all equations and the identification of the parameters is virtually the same as in the example above.

A very large collection of experimental data for many binary systems, from many sources, fit to the activity coefficient models is incorporated within Aspen Plus. An example of Wilson-2 parameters for an ethanol–water system is shown in Figure 6.3. These are available for user development of Aspen Plus simulations. Note that the temperature range of applicability is shown. These may not, however, satisfy requirements for many applications. For example, temperatures and pressures may not be appropriate to the modeling needs of specific applications. An important task for a user is to locate suitable data and customize (i.e., fit) the data to a model that lends itself to the application. If the required binary data are not available, the generalized correlation, Unifac, can be used to estimate the VLE data, which may then be fit to any of the activity coefficient equations. In such a situation the accuracy of a simulation may be compromised, and caution is advised.

6.3   BASIC IDEAS OF REGRESSION

Frequently, engineers and scientists are faced with sets of data that are proposed to fit a linear equation. The typical first pass at fitting the data is an eyeball plot on linear coordinates. This plot will usually differ when several persons attempt to plot the same data, and for the model

$$y = ax + b \tag{6.17}$$

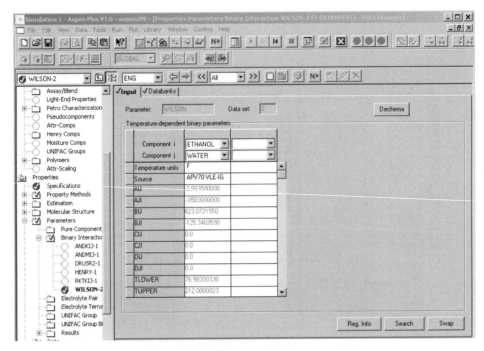

Figure 6.3   Wilson equation parameters.

the values of the slope, a, and intercept, b, will differ. This problem is solved using the well-known method of least squares, which works in the folllowing manner. An objective function,

$$\psi = \sum_i^{n\text{data}} (y_{\text{calc}} - ax_{\text{data}} - b)^2 \tag{6.18}$$

is a measure of the goodness of fit and is defined as the sum of squares of errors between the individual values of the dependent values of each of the y data points observed and the values of y calculated from the linear model, with trial values of a and b. The minimum of this function can be found by setting $\partial\psi/\partial a$ and $\partial\psi/\partial b$ equal to zero. This generates a simultaneous set of linear equations which can be solved for the values of a and b that best fit the data.

The problem of fitting data to the parameters of a nonlinear activity coefficient model is similar to the above. In Aspen Plus an objective function must be selected from several alternatives, given in Table 6.1. The maximum likelihood objective function is minimized directly subject to the constraints imposed by the applicable thermodynamic relationships using nonlinear programming methods such as those described by Edgar et al. (2001). All the remaining objective functions may be minimized by solving the nonlinear equations that arise by setting the partial derivatives of the objective function relative to the model parameters equal to zero.

TABLE 6.1 Objective Functions, Aspen Plus Version 7.0

Objective Function	Formed By:
Maximum likelihood	All measured variables
Ordinary least squares	Pressure and vapor compositions for isothermal VLE data
	Temperature and vapor compositions for isobaric VLE data
Barker's method	Pressure only
Modified Barker's method	Pressure and vapor compositions
Activity coefficients	Activity coefficients
Equilibrium constants	Equilibrium constants (K values)
Relative volatility	Relative volatility relative to first component

The choice of objective function depends on the type are data that are available. For vapor–liquid equilibrium the data may be of the type $T-P-x-y$ or $T-P-x$, where T is temperature, P is pressure, and x and y are liquid- and vapor-phase mole fractions, respectively. $T-P-x-y$ data can be used with all objective functions except Barker's method, which is used with $T-P-x$ data only. Selection of an objective function may be found at Propereties/Regression by choosing New regression and the tab Algorithm.

An example of an objective function suitable for use with $T-P-x-y$ data is Activity Coefficients:

$$\psi = \sum_{j=1}^{n\text{data}} \sum_{i=1}^{n\text{comp}} \left(\frac{\gamma_i^{\text{obs}} - \gamma_i^{\text{calc}}}{\gamma_i^{\text{obs}}} \right)^2 \tag{6.19}$$

where the experimental activity coefficients are calculated using

$$\gamma_i^{\text{obs}} = \frac{y_i^j \phi_i(y_1^j, y_2^j P^j, T^j) P^j}{x_i^j P_i^V(T^{oj})} \tag{6.20}$$

the experimental values and trial values of the activity coefficient parameters. Note the scaling of the activity coefficients in the objective function by the denominator, γ_i^{obs}:

When VLE data are of the type $T-P-x$ the Barker objective function

$$\psi = \sum_{j=1}^{n\text{data}} (P_j - P_j^{\text{calc}})^2 \tag{6.21}$$

is indicated. Here P_j^{calc} is calculated using the trial values of the activity coefficient equation parameters and the experimental values

$$P_j^{\text{calc}} = \gamma_1^j x_1^j P v_1^j + \gamma_2^j x_2^j P v_2^j \tag{6.22}$$

Fitting liquid–liquid equilibrium data (i.e., $T-x^{\text{I}}-x^{\text{II}}$, where x^{I} and x^{II} refer to the composition of the two liquid phases in equilibrium) can be done with the least squares or the maximum likelihood objective function.

6.4 MATHEMATICS OF REGRESSION

Two approaches for regressing data are available in Aspen Plus, minimizing an objective function by solving simultaneous nonlinear equations similar to linear regression and minimizing an objective function subject to the constraints of equilibrium. The partial derivatives of the objective functions with respect to the parameters are calculated numerically by perturbing each of the parameters in turn by a small value, Δa, a small fraction of the parameter, while keeping the remaining parameters constant. Using the perturbed and current values of the objective function, partial derivatives are calculated using

$$\frac{\partial \psi}{\partial a} = \frac{\psi_{a+\Delta a} - \psi_a}{\Delta a} \tag{6.23}$$

6.4.1 Newton–Raphson Method for Solution of Nonlinear Equations

The generalized Newton–Raphson method involves the arrangement of a set of n equations to be solved in the format

$$f_1(x_1, x_2, \ldots, x_n) = 0$$

$$f_2(x_1, x_2, \ldots, x_n) = 0$$

$$\vdots$$

$$f_n(x_1, x_2, \ldots, x_n) = 0$$

Each of the functions is expressed as a first-order Taylor's series about a point $\mathbf{X}^k$ to represent the functions at a point $\mathbf{X}^{k+1}$ in the vicinity of $\mathbf{X}^k$:

$$f_1(\mathbf{X}^{k+1}) = f_1(\mathbf{X}^k) + \left.\frac{\partial f_1}{\partial x_1}\right|_{X^k} (x_1^{k+1} - x_1^k) + \left.\frac{\partial f_1}{\partial x_2}\right|_{X^k} (x_2^{k+1} - x_2^k) + \cdots + \left.\frac{\partial f_1}{\partial x_n}\right|_{X^k} (x_n^{k+1} - x_n^k)$$

$$f_2(\mathbf{X}^{k+1}) = f_2(\mathbf{X}^k) + \left.\frac{\partial f_2}{\partial x_1}\right|_{X^k} (x_1^{k+1} - x_1^k) + \left.\frac{\partial f_2}{\partial x_2}\right|_{X^k} (x_2^{k+1} - x_2^k) + \cdots + \left.\frac{\partial f_2}{\partial x_n}\right|_{X^k} (x_n^{k+1} - x_n^k)$$

$$\vdots$$

$$f_n(\mathbf{X}^{k+1}) = f_n(\mathbf{X}^k) + \left.\frac{\partial f_n}{\partial x_1}\right|_{X^k} (x_1^{k+1} - x_1^k) + \left.\frac{\partial f_n}{\partial x_2}\right|_{X^k} (x_2^{k+1} - x_2^k) + \cdots + \left.\frac{\partial f_n}{\partial x_n}\right|_{X^k} (x_n^{k+1} - x_n^k)$$

Hypothesizing that all functions are zero at the point $\mathbf{X}^{k+1}$, evaluating all derivatives at the point $\mathbf{X}^k$, and changing the nomenclature such that

$$\Delta x_i = x_i^{k+1} - x_i^k$$

produces the following linear set of equations, which can be solved for the vector of Δx.

$$f_1(\mathbf{X}^{k+1}) = 0 = f_1(\mathbf{X}^k) + \left.\frac{\partial f_1}{\partial x_1}\right|_{X^k} \Delta x_1^k + \left.\frac{\partial f_1}{\partial x_2}\right|_{X^k} \Delta x_2^k + \cdots + \left.\frac{\partial f_1}{\partial x_n}\right|_{X^k} \Delta x_n^k$$

$$f_2(\mathbf{X}^{k+1}) = 0 = f_2(\mathbf{X}^k) + \left.\frac{\partial f_2}{\partial x_1}\right|_{X^k} \Delta x_1^k + \left.\frac{\partial f_2}{\partial x_2}\right|_{X^k} \Delta x_2^k + \cdots + \left.\frac{\partial f_2}{\partial x_n}\right|_{X^k} \Delta x_n^k$$

$$\vdots$$

$$f_n(\mathbf{X}^{k+1}) = 0 = f_n(\mathbf{X}^k) + \left.\frac{\partial f_n}{\partial x_1}\right|_{X^k} \Delta x_1^k + \left.\frac{\partial f_n}{\partial x_2}\right|_{X^k} \Delta x_2^k + \cdots + \left.\frac{\partial f_n}{\partial x_n}\right|_{X^k} \Delta x_n^k$$

The algorithm for solving a set of nonlinear simultaneous equations by the Newton–Raphson method is given below. If the equations to be solved are not analytical, it is necessary to calculate derivatives numerically.

1. Guess $\mathbf{X}^k$.
2. Evaluate all $f_i(\mathbf{X}^k)$.
3. If all $|f_i(\mathbf{X}^k)| < \varepsilon$, the problem is solved.
4. Calculate all partial derivatives at $\mathbf{X}^k$.
5. Solve the matrix–vector equations

$$\mathbf{P}\Delta\mathbf{X} = -\mathbf{F} \text{ for } \Delta\mathbf{X} \qquad \Delta\mathbf{X} = \mathbf{P}^{-1}(-\mathbf{F})$$

where $\mathbf{P}$ is the matrix of partial derivatives, $\Delta\mathbf{X}$ is the vector of x moves, and $\mathbf{F}$ is the vector of function values.

6. For all x_i, calculate

$$x_i^{k+1} = x_i^k + \Delta x_i$$

7. Return to step 2, substituting $\mathbf{X}^{k+1}$ for $\mathbf{X}^k$.

6.4.2 Direct Optimization of an Objective Function

The default regression method in Aspen Plus is maximum likelihood. In this method the objective function is given by

$$\psi = \sum_i^{n\text{data}} \left[\frac{(T_i^e - T_i^m)^2}{\sigma_{T,i}^2} + \frac{(P_i^e - P_i^m)^2}{\sigma_{P,i}^2} + \frac{(x_i^e - x_i^m)^2}{\sigma_{x,i}^2} + \frac{(y_i^e - y_i^m)^2}{\sigma_{y,i}^2} \right] \qquad (6.24)$$

where the superscripts e and m refer to estimated and measured values. For each point in the set of data the following constraints apply. For simplicity, Poynting correction is ignored; then for each component one may write

$$y_i \phi_i^V P - \gamma_i x_i f_i^0 = 0 \qquad (6.25)$$

$$\gamma_i = \gamma(T, x) \qquad (6.26)$$

$$\phi_i^V = \phi^V(T, P, y) \qquad (6.27)$$

and

$$\sum_{i}^{ncomp} x_i - 1 = 0 \tag{6.28}$$

$$\sum_{i}^{ncomp} y_i - 1 = 0 \tag{6.29}$$

Aspen Plus documentation does not provide information on the details of the implementation of the minimization procedure; however, the papers of Anderson et al. (1976) and Fabries and Renon (1976) describe possibilities.

6.5 PRACTICAL ASPECTS OF REGRESSION OF VLE OR LLE DATA

The first question that arises when activity coefficient equations are employed is whether to use a two- or four-parameter approach. With the NRTL equation it would be either three or five. Aspen Plus recommends that the NRTL c parameter be fixed according to the following rules:

- 0.2 for saturated hydrocarbons with polar nonassociated liquids and systems that exhibit liquid–liquid immiscibility
- 0.3 for nonpolar substances; nonpolar with polar nonassociated liquids; small deviations from ideality
- 0.47 for strongly self-associated substances with nonpolar substances

When using two parameters, a question arises as to the selection of the constant $a_{i,j}$ or the temperature-dependent $b_{i,j}$ version of the parameters. Prausnitz et al. (1999) discuss this point and no clear advice is available; however, experience has shown that the difference between the two approaches in quality of fit is very small. Often, the difference between the two- and four-parameter approach is also very small. When regressing four parameters simultaneously, steps taken during the regression are related to the current values. It is important to note that relative to the $b_{i,j}$ values, the $a_{i,j}$ values vary sometimes by two orders of magnitude. Aspen Plus provides the ability to assign a scaling factor to any parameter so that step sizes during the regression are not dominated by a particular parameter.

In most cases of nonlinear data regression, more than one set of values that satisfy the equations can be found. It is imperative that results be checked to assure that the results produce calculated values of the independent values that fit the data consistent with the requirements of the simulation for which the results will be employed. In many cases the values calculated by the regression will depend on the starting values of the parameters. In some cases a regression that employs Aspen Plus's default values will suffice.

6.5.1 Regression of VLE Data

A useful technique for providing starting values for vapor–liquid equilibrium regression using $T-P-x-y$ data is to estimate the infinite dilution activity coefficients by calculating activity coefficients using equation (6.18), plotting versus x_1 and extrapolating to

TABLE 6.2 Infinite Dilution Activity Coefficients for Various Equations

Equation	$\ln \gamma_1^\infty$ and $\ln \gamma_2^\infty$
Van Laar	$\ln \gamma_1^\infty = A_{12}$; $\ln \gamma_2^\infty = A_{21}$
Wilson	$\ln \gamma_1^\infty = 1 - \ln \Lambda_{12} - \Lambda_{21}$; $\ln \gamma_2^\infty = 1 - \ln \Lambda_{21} - \Lambda_{12}$
NRTL	$\ln \gamma_1^\infty = \tau_{21} + \tau_{12} \exp(-\alpha_{12}\tau_{12})$; $\ln \gamma_2^\infty = \tau_{12} + \tau_{21} \exp(-\alpha_{12}\tau_{21})$
Uniquac	$\ln \gamma_1^\infty = \ln(r_1/r_2) + [5\ln(q_1 r_2/q_2 r_1) - \ln \tau_{21} + 1 - \tau_{12}] + I_1 - (r_1/r_2)I_2$
	$\ln \gamma_2^\infty = \ln(r_2/r_1) + [5\ln(q_2 r_1/q_1 r_2) - \ln \tau_{12} + 1 - \tau_{21}] + I_2 - (r_2/r_1)I_1$

the 0.0 and 1.0. values of x_1. These are related to the parameters by various algebraic expressions, as shown in Table 6.2.

The adjustable parameters are A_{ij} for van Laar, Λ_{ij} for Wilson, τ_{ij} and α_{12} for NRTL, and τ_{ij} for Uniquac. The r, q, and I parameters are pure component values which can be found in the book by Poling et al. (2000).

An alternative approach is to recognize that a single data point yields two equilibrium equations with the parameters a_{ij} as unknowns. Solution of these equations yields good estimates of the a_{ij} for a reasonable data point.

Both of these approaches can be implemented easily using the Aspen Plus regression system. The former approach would involve using the entire data set for the regression with the initial values set to zero. If this produces good results, nothing more need be done; if not, the experimental activity coefficients are listed under regression/results/tab profiles and can be displayed with Aspen Plus's plotting capability and extrapolated by eyeball.

The latter approach can also be implemented with Aspen Plus by regressing a single data point. Subsequently, the regression with the full data set can be done with the estimated values of the a_{ij} pair as starting values. If one decides to regress the b_{ij} pair, an initial value can be obtained by noting that $a_{ij} = b_{ij}/T$ in the absence of a four-parameter fit.

Aspen Plus provides a very useful Plotting Wizard which has many built-in plots related to regression of equilibrium data (see Figure 6.4, as well as the ability to select which variables to plot from a variety of Aspen Plus data sources. The Wizard is invoked by selecting Plot from the main display menu. All results shown may be obtained by selecting Properties/Regression/R1 with tab Profiles and may be plotted by selecting the column to be plotted and associating it with the desired axis available in the Plot menu. Figure 6.5 shows an example of a gamma(1) and gamma(2) versus x_1 plot.

As an example, Table 6.3 gives the T–P–x–y data for the system cyclohexane (CH)–isopropyl alcohol (IPA) of Nagata (1963). A plot of the activity coefficients of both components versus the mole fraction of component 1 (CH) is shown in Figure 6.5. Extrapolation of the curves produces infinite dilution activity coefficients of about 4.8 for CH and 9.5 for IPA.

Applying the infinite dilution form of the van Laar equation to the estimated infinite dilution coefficients yields 1.57 and 2.25 for a_{12} and a_{21}, respectively. Calculating b_{12} and b_{21} and noting that the original van Laar parameter A_{12} is modified to

$$A_{12} = a_{12} + \frac{b_{12}}{T}$$

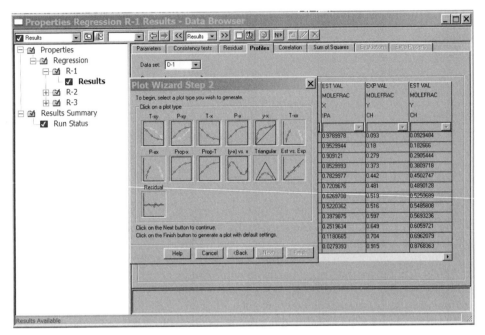

Figure 6.4   Aspen Plus plot wizard.

TABLE 6.3   Vapor–Liquid Equilibrium
Cyclohexane (1)–Isopropyl Alcohol (2)

Temperature (°C)	Pressure (mmHg)	x_1	y_1
80.29	760	0.021	0.093
78.1	760	0.047	0.18
78.85	760	0.091	0.279
73.8	760	0.147	0.373
72.13	760	0.217	0.442
70.88	760	0.279	0.481
70.13	760	0.373	0.519
69.99	760	0.478	0.561
69.56	760	0.602	0.597
69.79	760	0.748	0.649
70.99	760	0.882	0.704
74.61	760	0.972	0.815

then either $A_{12} = a_{12}$ or $A_{12} = b_{12}/T$ when two parameters are used. Then for constant A_{12}, an initial estimate for $b_{12} = a_{12}T$ and similarly for b_{21}. For this example,

$$b_{12} = 1.57(273.1 + 80.3) = 530.1$$
$$b_{21} = 2.25(273.1 + 74.6) = 782.3$$

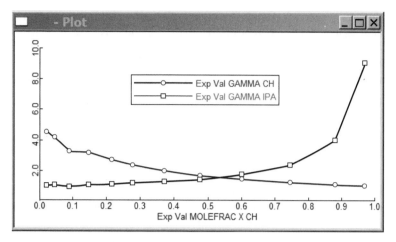

Figure 6.5 Activity coefficient plot.

This is illustrated from Aspen Plus regression runs, which give the following results. When $b_{ij} = 0$:

$$a_{12} = 1.435$$
$$a_{21} = 2.298$$

When $a_{ij} = 0$:

$$b_{12} = 498.8$$
$$b_{21} = 791.1$$

Virtually identical results were obtained with least squares and maximum likelihood regressions.

When setting up multiple regressions in one run, care must be taken to set the unused parameters equal to zero because Aspen Plus will use older or stored values in the regression. The ideas illustrated above are appropriate for any choice of activity coefficient equation. Details of the regression runs above are available at Examples/chipavanLaar and Examples/chipwilson.

6.5.2 Regression of LLE Data

When fitting liquid–liquid equilibrium data the situation is somewhat different. Most extraction processes are isothermal; therefore, if binary mutual solubility data are available, consistent with ternary (or higher) LLE data at the same temperature, the $a_{i,j}$ can be obtained by using one data point for regression. Since each data point can generate two equations,

$$(\gamma_1 x_1)^{\mathrm{I}} = (\gamma_1 x_1)^{\mathrm{II}} \tag{6.30}$$

an exact solution is possible.

With suitable data it is possible to estimate the infinite dilution activity coefficients by assuming that when the composition of a component is near a mole fraction of 1.0, its activity coefficient is 1.0. For example, in the system A–B the composition of component A in phase I, x_1^I, is 0.002 and the composition of component B in phase II, x_2^{II}, is 0.01. Applying equation (6.30) for each component yields, approximately at infinity, $\gamma_1^I = 495$ and $\gamma_2^{II} = 998$.

If the $b_{i \cdot j}$ parameters are required because the application involves a variable temperature, several points must be regressed, although an initial value can be obtained from one data point using one of the methods above.

When regressing ternary systems, two possibilities exist: (1) fit the two solvents with mutual solubility data and fit the four remaining parameters with ternary data while holding the two previously determined parameters constant; or (2) fit all six parameters with ternary data. Using the ternary data of Sugi and Katayama (1978) shown in Table 6.4, we look next at examples of both methods.

TABLE 6.4   LLE Data for the System n-Hexane (2)–Ethanol (2)–Acetonitrile (3)

Temperature °K	Mole Fraction Phase I		Mole Fraction Phase II	
	N-Hexane	Ethanol	N-Hexane	Ethanol
313.15	0.8831	0.0166	0.0968	0.0879
313.15	0.8674	0.0251	0.1003	0.1299
313.15	0.8546	0.0332	0.1054	0.1622
313.15	0.8372	0.0433	0.125	0.1942
313.15	0.8101	0.0567	0.1367	0.2169
313.15	0.7821	0.0772	0.1522	0.2331
313.15	0.7272	0.1086	0.2053	0.2695
313.15	0.6912	0.1269	0.2249	0.2748

First Method   An estimate of the mutual solubility data is made by extrapolating a plot of the data from Figure 6.6, resulting in $x_1^I = 0.9$ and $x_1^{II} = 0.09$. The estimates calculated for the parameters are

$$b_{13} = -553.43076$$
$$b_{31} = -18.956075$$

Holding b_{13} and b_{31} constant and fitting the remaining four parameters to the experimental data yields the following results:

$$b_{12} = 10.801713$$
$$b_{21} = -169.74093$$
$$b_{23} = 215.31305$$
$$b_{32} = -178.56357$$

Second Method   Parameters obtained by calculating the values of six parameters using one data point are

$$b_{12} = -494.10342$$

$$b_{12} = 95.2601793$$

$$b_{13} = -530.14039$$

$$b_{31} = -35.602627$$

$$b_{23} = -1.3683628$$

$$b_{32} = -253.9741$$

Parameters obtained using the full data set initialized with the values above are

$$b_{12} = 1892.22192$$

$$b_{12} = 136.837046$$

$$b_{13} = -461.75675$$

$$b_{31} = -68.070246$$

$$b_{23} = 12.3152652$$

$$b_{32} = 2285.7669$$

All the methods described give passable results, and careful comparison of results with the composition range of the process would facilitate a choice of parameters. Aspen Plus's Plot Wizard offers several possibilities for viewing experimental and calculated values of the variables. An example is shown in Figure 6.6. Details of the Aspen Plus regressions above are available at Examples/lle method one one point, Examples/lle method one two parameters fixed, Examples/lle method two one point, and Examples/lle method two.

6.6 WORKSHOPS

In preparation for the workshops it would be advisable to read the Aspen Plus documentation dealing with the activity coefficient models van Laar, Wilson, NRTL, Uniquac,

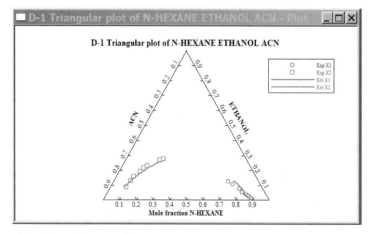

Figure 6.6 Triangular plot of LLE data of an *n*-hexane–ethanol–acetonitrile system.

Unifac, and Unifac-LL. Note the similarity between Uniquac and Unifac. Note the difference in application between Unifac and Unifac-LL.

Workshop 6.1 Use Property Analysis to estimate the mutual solubility of a toluene–water system at temperatures varying from 32 to 86°F at 14.696 psi. Select the property option Unifac-LL. The mutual solubility of the toluene–water system at 86°F and 14.696 psi is 0.999894 and 0.000237 mole fraction water in each phase, respectively. Use engineering (ENG) units. Compare to the results estimated.

Workshop 6.2a Fit the following single data point of mutual solubility for a toluene–water system to the two-parameter form of the van Laar equation. Use the parameters $a_{i,j}$ and $a_{j,i}$. At 25°C and 14.696 psi the solubility of water expressed as mole fractions in each phase is 0.999894 and 0.000237, respectively.

Workshop 6.2b Fit the toluene–water solubility data from the LLE Dechema data collection in Table 6.5 to the van Laar equation using the $a_{i,j}$ and $a_{j,i}$ parameters. All solubility data are at 14.696 psi and expressed in mole fractions of toluene. Use engineering units.
 What are your observations?

TABLE 6.5 Toluene–Water Mutual Solubility Data

Temperature (°C)	Mole Fraction x^{I}	Mole Fraction x^{II}
0	0.000142	0.99891
10	0.000128	0.99844
20	0.000113	0.99784
25	0.000106	0.99763

Workshop 6.2c Repeat Workshop 6.2b using $b_{i,j}$ and $b_{j,i}$ parameters with the van Laar equation. What are your observations?

Workshop 6.3a Repeat Workshop 6.2b using $b_{i,j}$ and $b_{j,i}$ parameters with the Uniquac equation. What are your observations?

Workshop 6.3b Using the data of Washburn et al. (1939) for the water–ethanol–toluene system in Table 6.6, fit the data to the Uniquac equation with the $b_{i,j}$ parameters determined from the single mutual solubility data point of Workshop 6.3a held constant. Vary the $b_{i,j}$ parameters for the remaining binary pairs.

Workshop 6.3c Repeat Workshop 6.3b varying all six $b_{i,j}$ parameters? Compare the results to those of Workshop 6.3b. Which result should be used?

Workshop 6.4a Fit the data of Suska et al. (1970) given in Table 6.7 to the van Laar, Wilson, NRTL, and Uniquac equations. Use the a_{ij} parameters for all equations. Additionally, use the c_{ij} parameter when fitting the NRTL equation. Select and fix the appropriate value of the $c_{I,j}$ parameter. Compare the four results.

TABLE 6.6   Liquid–Liquid Equilibrium Data for a Water–Ethanol–Toluene System

Temperature (°C)	Pressure (psi)	Phase-W Mole Fraction		Phase-T Mole Fraction	
		Water	Ethanol	Water	Ethanol
25	14.696	0.95320	0.04621	0.00356	0.00390
25	14.696	0.91360	0.08538	0.00506	0.01385
25	14.696	0.87856	0.11990	0.00703	0.02552
25	14.696	0.82021	0.17712	0.01150	0.05210
25	14.696	0.77323	0.22240	0.01960	0.08849
25	14.696	0.72620	0.26547	0.02007	0.09905
25	14.696	0.69166	0.29572	0.02425	0.11970
25	14.696	0.64420	0.33500	0.03111	0.14779
25	14.696	0.59615	0.37190	0.03719	0.16672

TABLE 6.7   Vapor–Liquid Equilibrium for a Tertiary Butanol (1)–Water (2) System

Temperature (°C)	Pressure (mmHg)	Liquid Mole Fraction: Tertiary Butanol	Vapor Mole Fraction: Tertiary Butanol
86.50	760	0.0190	0.4040
83.80	760	0.0270	0.4630
82.00	760	0.0430	0.4840
81.70	760	0.0630	0.4890
81.40	760	0.1190	0.4990
81.10	760	0.1670	0.5070
81.00	760	0.2340	0.5190
80.90	760	0.3160	0.5340
80.65	760	0.3690	0.5440
80.50	760	0.3880	0.5460
80.30	760	0.4670	0.5700
80.00	760	0.5290	0.5890
80.00	760	0.6090	0.6220
80.00	760	0.6170	0.6280
80.00	760	0.6380	0.6410
80.10	760	0.6710	0.6960
80.20	760	0.7550	0.7160
80.30	760	0.8280	0.7850
80.90	760	0.8740	0.8330
81.25	760	0.9120	0.8770
81.80	760	0.9410	0.9050

Workshop Notes

Workshop 6.1   In implementing this workshop an approach similar to one that would be used in a laboratory is advisable. For example, if equal quantities of toluene and water were placed in a beaker, mixed thoroughly, and permitted to stand for a sufficient time to equilibrate, the phases would form and could be sampled. Chemical analysis of the samples would yield the result desired.

Under the subject Analysis and Tab System, select "generate points along a flash curve" and valid phases "vapor–liquid–liquid" even though there will be no vapor. Set up a mixture for equilibration as described using the following procedure. Under the tab Variable, select pressure as a fixed state variable and temperature as the adjusted variable. Select and highlight temperature and the Range/List button will become available to enter temperature data.

Workshop 6.2a   The fit is exact because a single mutual solubility point is described by two equilibrium equations like equation (6.30) which will contain as unknowns only the two parameters of the activity coefficient equation.

Workshop 6.2b and 6.2c   As expected, use of the a_{ij} parameter produces values of solubility that are independent of temperature; however, even when using the b_{ij} parameters, the data are not well represented.

Workshop 6.3a   An exact solution is obtained as in Workshop 6.2a.

Workshop 6.3b and 6.3c   When viewing results using the Plot Wizard's est vs. exp plot, there are small differences between the two approaches, but this is not a universal result.

Use of the mutual solubility data point as in Workshop 6.3b establishes the location of the phase envelope termini (on a ternary diagram) on the toluene–water edge of the diagram and will force the composition at very dilute phases to approach the experimental mutual solubility. If the mutual solubility data are not consistent with the other experimental tie lines, the phase envelope in that area of the phase diagram will not appear smooth. If the data are fit as in Workshop 6.3c, the phase envelope will be smooth, but the mutual solubility predicted from the ternary data may not conform to the experimental mutual solubility. Because of this potential problem, an Aspen Plus user must be cognizant of the expected compositions of the simulation for which the regression is being prepared and make the appropriate choice when the data are regressed.

Workshop 6.4a   The fit of the data to the van Laar equation suggests that there is a heterogeneous region that lies between 0.1 to 0.25 mole fraction TbOH. A careful look at the experimental data shows that this may be so, but it is a judgment call. More experimental data may be needed. Extreme caution is advised.

The Wilson equation fit by its nature cannot have a heterogeneous region and is otherwise satisfactory. The Uniquac fit exhibits behavior similar to that of van Laar. The NRTL fit with the c parameter $= 0.2$ exhibits similar behavior to that of van Laar.

Workshop 6.4b   The NRTL fit with the c parameter $= 0.3$ exhibits behavior similar to that of van Laar.

REFERENCES

Anderson, T. F., Abrams, D. D., and Grens, E. A., II, *AIChE. J.*, 24(1), 20–39 (1976).
Aspen Plus version 7.0, Physical Properties System documentation.

Edgar, T. F., Himmelblau, D. M., and Lasdon, L. S. *Optimization of Chemical Processes*, 2nd ed., McGraw-Hill, New York, 2001.

Fabries, J. F., and Renon, H., *AIChE. J.*, 21(4), 735–742 (1976).

Knapp, H., Döring, R., Oellrich, L., Plöcker, U., Prausnitz, J. M., Langhorst, R., and Zeck, S. Chemistry Data Series, Vol. VI; *VLE for Mixtures of Low Boiling Substances*, Dechema, Frankfort am Main, Germany, 1982.

Nagata, I., *Men. Proc. Technol. Kanazawa Univ.*, 3(I), 1 (1963).

Poling, B. E., Prausnitz, J. M., and O'Connell, J. P., *The Properties of Gases and Liquids*, 5th ed., McGraw-Hill, New york, 2000.

Prausnitz, J. M., Lichtenthaler, R. N., and de Avezedo, E. G., *Molecular Thermodynamics of Fluid-Phase Equilibria*, 3rd ed., Prentice Hall, Upper Saddle River, NJ, 1999, p. 253.

Sugi, H., and Katayama, T., *J. Chem. Eng. Jpn.*, 11(7), 167–172 (1978).

Suska, J., Holub, R., Vonka, P., and Pick, J., *Collect. Czech. Chem. Commun.*, 35, 3851 (1970).

Washburn, R. E., Beguin, A. E., and Beckord, O. C., *J. Am. Chem. Soc.*, 61, 1694 (1939).

CHAPTER SEVEN

FLASHES AND DECANTER

Aspen Plus's model library contains two different rigorous flash blocks that solve the appropriate material, energy balance, and equilibrium equations. The Flash2 block is designed to produce a single vapor phase and a single liquid phase that are in equilibrium when the flash conditions are specified such that a multicomponent mixture's state is in the two-phase region. Similarly, the Flash3 block is designed to produce one vapor phase and two liquid phases in equilibrium for suitably specified process conditions. The Flash3 block is also capable of solving a liquid–liquid equilibrium problem under conditions where no vapor is produced. A similar block is Decanter, which is designed to produce two liquid phases in equilibrium in the absence of a vapor phase. The Flash2, Flash3, and Decanter blocks can be found in the model library under the Separators tab.

7.1 Flash2 BLOCK

Figure 7.1 is a graphical depiction of a Flash2 block. The nomenclature used is as follows: F represents the total feed flow, in moles/time; f_i the flow of component i in the feed, in moles/time; V the total vapor flow, in moles/time; v_i the flow of component i in the vapor, in moles/time; L the total liquid flow, in moles/time; l_i the flow of component i in the liquid, in moles/time; and n the number of components. Including the flash temperature T_f and pressure P_f results in $2n + 2$ independent variables given the feed state. These are equilibrium temperature; flash pressure; two total flows, L and V; and $2(n-1)$ component flows; or alternatively, $2n$ component flows, excluding the total flows. Mole fractions are calculated from the independent

Teach Yourself the Basics of Aspen Plus™   By Ralph Schefflan
Copyright © 2011 John Wiley & Sons, Inc.

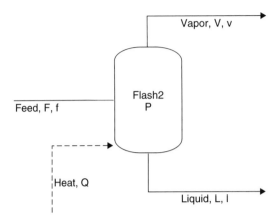

Figure 7.1   Flash2 model.

variables by an equation such as

$$y_i = \frac{v_i}{\sum_{j=1}^{n} v_j} \tag{7.1}$$

where y_i is the mole fraction of component i in the vapor.

The applicable material balances are n componential equations such as

$$f_i - v_i - l_i = 0 \tag{7.2}$$

or alternatively, $n - 1$ equations such as equation (7.2) and one overall material balance given by

$$F - V - L = 0 \tag{7.3}$$

Additionally, n equilibrium equations which describe the equality of the fugacities of components in each phase are required. When the liquid fugacity is represented by an equation of state, where ϕ_i^L and ϕ_i^V are the fugacity coefficients of component i in the liquid and vapor phases, respectively, the result is

$$y_i \phi_i^V - x_i \phi_i^L = 0 \tag{7.4a}$$

When the liquid fugacity is represented by an activity coefficient equation, where γ_i is the activity coefficient of component i and the vapor phase is represented by an equation of state

$$y_i \phi_i^V P - \gamma_i x_i p_i^v = 0 \tag{7.4b}$$

where p_i^v is the vapor pressure of component i, results. For the sake of simplicity, the Poynting correction (see Prausnitz et al., 1999), which has a contribution only for very light components, has been omitted.

An overall energy balance, where h with a suitable subscript represents the enthalpy of the feed, liquid, and vapor, respectively, and Q represents the heat added, given by

$$h_F F + Q - h_L L - h_V V = 0 \tag{7.5}$$

completes the set of $2n + 1$ equations.

The Flash2 block permits the specification of two of the four possibilities: flash temperature, flash pressure, heat required, and the fraction of the feed vaporized, except the combination Q and V/F. Various permutations of the specifications require other modifications of the equations and the list of unknowns. For example, if V/F is specified, equation (7.5) is replaced by

$$\left(\frac{V}{F}\right)_{\text{specified}} - \frac{\sum_{j=1}^{n} v_j}{\sum_{j=1}^{n} f_j} = 0 \tag{7.6}$$

If Q, the heat added or removed, and the flash pressure are specified, the $2n$ component flows and the flash temperature can be calculated. If the flash temperature and pressure are specified, the sequence of the solution of the equations changes. Equations (7.1) through (7.4) are solved simultaneously, after which equation (7.5) is solved for Q.

An example of the use of a Flash2 block is given at Chapter Seven Examples/ Flash2Example. An important aspect of the setup, which is applicable to all simulations, is the selection of the data bank from which the activity coefficient or equation of state parameters were derived. Figure 7.2 shows that the binary interaction Wilson

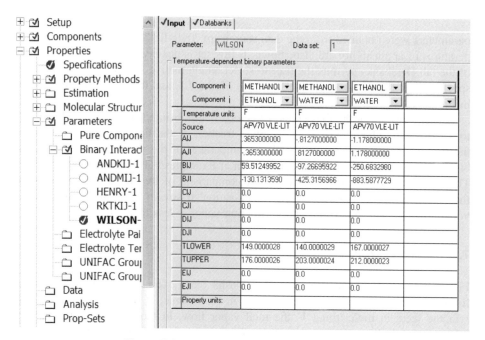

Figure 7.2 Flash2 example Wilson parameters.

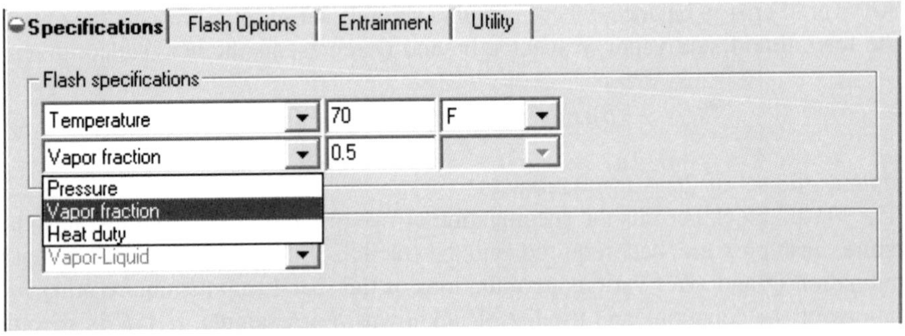

Figure 7.3   Flash2 example specifications.

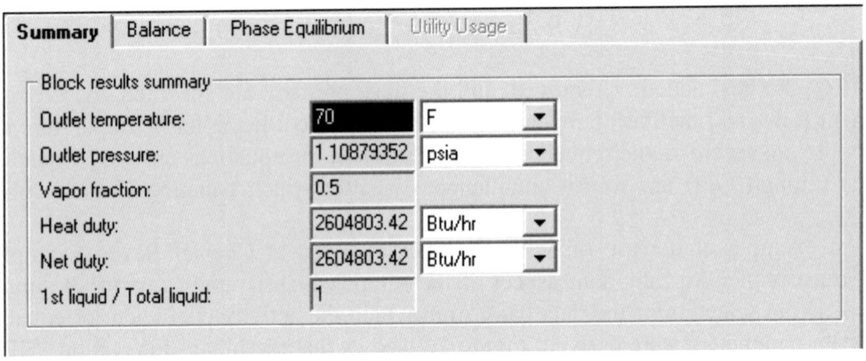

Figure 7.4   Flash2 example results.

parameters were obtained from the data source apv70vle-lit. The upper and lower temperatures of applicability are also shown. If this set is deemed inappropriate, selection of the tab Databanks permits the user to select an alternative source of data. The input specifications for this example are shown in Figure 7.3, and the flash results are given in Figure 7.4.

A useful feature of the Flash2 and some other equilibrium stage blocks is the ability to specify entrainment; that is, a specified fraction of the liquid phase is carried by the flashing vapor into the overhead. A comparison of the performance of a Flash2 block with and without entrainment is given at Examples/Flash2EntrainmentExample. The flowsheet is shown in Figure 7.5. A comparison of the overheads produced by the flash is given in Figure 7.6. The composition of stream 22 is lower in the more volatile component, methanol, and its fraction vapor is less than 1.0.

7.2   Flash3 BLOCK

Figure 7.7 is a graphical depiction of a Flash3 block. The nomenclature used is as follows: F represents the total feed flow, in moles/time; f_i the flow of component i in the feed, in moles/time; V the total vapor flow, in moles/time; v_i the flow of component i in the vapor, in moles/time; L^1 the total flow of the first liquid phase, in moles/time; l_i^l, the flow of component i in the liquid, in moles/time; L^2 the total flow

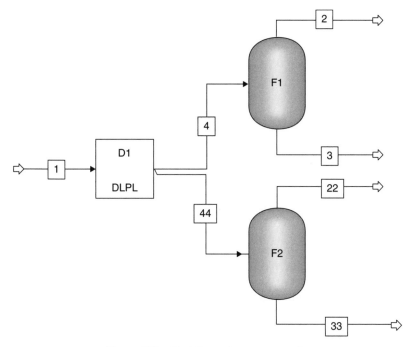

Figure 7.5   Flash2 entrainment example.

Material	Heat	Load	Work	Vol.% Curves	Wt. % Curves	Petro. Curves	Poly. Curves

Display: Streams ▼   Format: CHEM_E ▼   Stream Table

	2 ▼	22 ▼	▼
Temperature F	186.0	186.0	
Pressure psia	20.00	20.00	
Vapor Frac	1.000	0.903	
Mole Flow lbmol/hr	48.257	53.432	
Mass Flow lb/hr	1332.485	1449.524	
Volume Flow cuft/hr	16718.876	16721.176	
Enthalpy MMBtu/hr	-4.384	-4.974	
Mole Flow lbmol/hr			
METHANOL	33.016	34.715	
WATER	15.241	18.717	

Figure 7.6   Flash2 entrainment results.

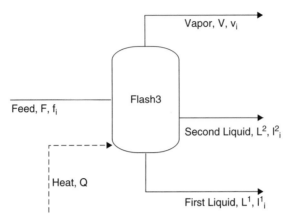

Figure 7.7   Flash3 model.

of the second liquid phase, in moles/time; and I_i^2 the flow of component i in the liquid, in moles/time. Including the flash temperature T_f and pressure P_f results in $3n + 2$ independent variables, where n is the number of components given the feed state. The variables consist of the equilibrium temperature, the flash pressure, and the three total flows (i.e., and L^1, L^2, and V), and $3(n-1)$ componential flows or, alternatively, $3n$ componential flows, excluding the total flows. Mole fractions are calculated from the independent variables by an equation such as equation (7.1).

The applicable material balances are n componential equations, such as

$$f_i - v_i - l_i^1 - l_i^2 = 0 \qquad (7.7)$$

or alternatively, $n-1$ equations such as equation (7.7) and one overall material balance, given by

$$F - V - L^1 - L^2 = 0 \qquad (7.8)$$

Additionally, $3n$ equilibrium equations which describe the equality of the fugacities of components in each phase can be written, but only $2n$ are independent. The vapor-phase fugacity is represented by an equation of state where the Φ_i^V are the fugacity coefficients of component i in the vapor phase, and the liquid phases are represented by an activity coefficient equation where γ_i^1 and γ_i^2 are the activity coefficients of component i and p_i^v is the vapor pressure of component i. Then any two of

$$\gamma_i^1 x_i^1 - \gamma_i^2 x_i^2 = 0 \qquad (7.9a)$$

$$y_i \phi_i^V P - \gamma_i^1, x_i^1 p_i^v = 0 \qquad (7.9b)$$

$$y_i \phi_i^V P - \gamma_i^2, x_i^2 p_i^v = 0 \qquad (7.9c)$$

apply. For the sake of simplicity the Poynting correction (see Poling et al., 2000), which has a contribution only for very light components, has been omitted from equations (7.9b) and (7.9c). The overall energy balance given by equation (7.5) completes a set of $3n + 1$ equations:

$$h_F F + Q - h_{L1} L^1 - h_{L2} L^2 - h_V V = 0 \tag{7.10}$$

The Flash3 block permits the specification of two of the four possibilities: flash temperature, flash pressure, heat required, and the fraction of the feed vaporized. Various permutations of the specifications require other modifications to the equations and the list of unknowns. For example, if Q, the heat added/removed and the flash pressure are specified, the $3n$ component flows and the flash temperature can be calculated.

When specifying the block it is important to identify the key component in the bottom takeoff (liquid 2) such that the block conforms with its physical behavior (i.e., the heaviest of the solvents should be selected). An example of the use of a Flash3 block is given at Examples/Flash3Example.

7.3 DECANTER BLOCK

Figure 7.8 is a graphical depiction of a Decanter block. The nomenclature used is as follows: F represents the total feed flow, in moles/time; f_i the flow of component i in the feed, in moles/time; L^1 the total flow of the first liquid phase, in moles/time; l_i^1 the flow of component i in the liquid, in moles/time; L^2 the total flow of the second liquid phase, in moles/time; and l_i^2 the flow of component i in the liquid, in moles/time. Including the decanter temperature T_d and pressure P_d results in $2n + 2$ independent variables, where n is the number of components given the feed state. The variables consist of the equilibrium temperature and the decanter pressure, the two total flows L^1 and L^2, and $2(n-1)$ componental flows or, alternatively, $2n$ componental flows, excluding the total flows. Mole fractions are calculated from the independent variables by an equation such as (7.1).

The applicable material balances are n componental equations, such as

$$f_i - l_i^1 - l_i^2 = 0 \tag{7.11}$$

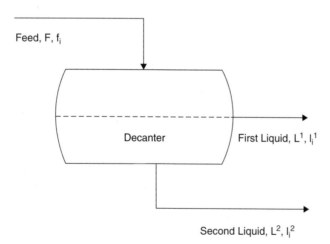

Figure 7.8 Decant model.

or alternatively, $n-1$ equations such as equation (7.11) and one overall material balance, given by

$$F - L^1 - L^2 = 0 \qquad (7.12)$$

Additionally, n equilibrium equations can be written which describe the equality of the fugacities of components in each phase, where γ_i^1 and γ_i^2 are the activity coefficients of component i and x_i^1 and x_i^2 are the mole fractions of component i in phases 1 and 2, respectively:

$$\gamma_i^1 x_i^1 - \gamma_i^2 x_i^2 = 0 \qquad (7.13)$$

An overall energy balance,

$$h_F F + Q - h_{L1} L^1 - h_{L2} L^2 = 0 \qquad (7.14)$$

completes a set of $2n + 1$ equations.

Most decantation processes are carried out isothermally, and the heat associated with solute moving between phases is usually negligible, thus eliminating the need for equation (7.14) and reducing the number of unknowns to $2n$ when the pressure is specified. This is reflected in the block's input forms, where the only allowable specifications are temperature and pressure. As in the Flash3 block, care must be taken to specify the key component in the second liquid product, which should be the heavier phase. The feed from Flash3Example was used to create an example of the use of a Decanter block, which may be found in Examples/DecanterExample. It would be useful to compare results of Flash3 and Decanter since both are carried out at the same conditions: 1 psia and 62°F. The Flash3 specification is $V/F = 0.1$, which produces a temperature of 61.9°F, which is close enough for a comparison. Figure 7.9 shows the data sources for the Uniquac parameters, which are LLE data compared to VLE data for Flash3. This is an example of a conundrum. Which data produce the correct answer? Both seem reasonable. Aspen Plus has solved the appropriate equations, but it is up to the user to select the correct result. Perhaps some laboratory work will be required.

The block also permits an alternative formulation of the basic equations as a Gibbs free energy minimization problem. This formulation and some methods of solution are described by Walas (1985). Additional details are provided in Chapter Ten.

An interesting aspect of the Decanter block is the possibility of its use to simulate a batch extraction process. The describing equations are identical to those of a continuous decanter if the flows per unit time are replaced with a batch charge as a feed and the batch removal of the two products. A common practice in industry is to place a solvent A containing a quantity of C, the product desired, into a vessel. Following this, the solvent B is charged to the vessel. After agitation the products are allowed to settle. This is the equivalent of feeding both streams as a mixture to the decanter, producing the products L^1 and L^2. Common industrial practice is to remove the equilibrium B stream, now containing extracted C, and recharging the vessel with fresh B. This process is easily simulated with several Decanter blocks in series, as shown in Figure 7.10. The combined feeds are repeatedly settled, the extract and raffinate removed, and the resulting product sent to the next stage. Each stage represents the same vessel one

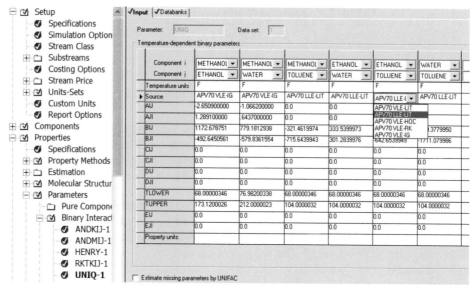

- ☑ Setup
  - ✿ Specifications
  - ✿ Simulation Option
  - ✿ Stream Class
  - ☐ Substreams
  - ✿ Costing Options
  - ☐ Stream Price
  - ☑ Units-Sets
  - ✿ Custom Units
  - ✿ Report Options
- ☑ Components
- ☑ Properties
  - ✿ Specifications
  - ☑ Property Methods
  - ☐ Estimation
  - ☑ Molecular Structur
  - ☑ Parameters
    - ☐ Pure Compone
    - ☑ Binary Interact
      - ✿ ANDKIJ-1
      - ✿ ANDMIJ-1
      - ✿ HENRY-1
      - ✿ RKTKIJ-1
      - ✿ **UNIQ-1**

✓Input ✓Databanks

Parameter: UNIQ   Data set: 1

Temperature-dependent binary parameters

Component i	METHANOL	METHANOL	METHANOL	ETHANOL	ETHANOL	WATER
Component j	ETHANOL	WATER	TOLUENE	WATER	TOLUENE	TOLUENE
Temperature units	F	F	F	F	F	F
▶ Source	APV70 VLE-IG	APV70 VLE-IG	APV70 LLE-LIT	APV70 LLE-LIT	APV70 LLE-I ▾	APV70 LLE-LIT
AIJ	-2.650900000	-1.066200000	0.0	0.0		
AJI	1.289100000	.6437000000	0.0	0.0		
BIJ	1172.678751	779.1812938	-321.4619974	333.5399973		.3779950
BJI	-492.6450561	-579.8361554	-715.6439943	301.2839976	-64Z.6539943	-1711.079986
CIJ	0.0	0.0	0.0	0.0	0.0	0.0
CJI	0.0	0.0	0.0	0.0	0.0	0.0
DIJ	0.0	0.0	0.0	0.0	0.0	0.0
DJI	0.0	0.0	0.0	0.0	0.0	0.0
TLOWER	68.00000346	76.98200338	68.00000346	68.00000346	68.00000346	68.00000346
TUPPER	173.1200026	212.0000023	104.0000032	104.0000032	104.0000032	104.0000032
EIJ	0.0	0.0	0.0	0.0	0.0	0.0
EJI	0.0	0.0	0.0	0.0	0.0	0.0
Property units:						

Dropdown options (col ETHANOL/TOLUENE Source): APV70 LLE-LIT, APV70 VLE-LIT, APV70 VLE-HOC, APV70 VLE-RK, APV70 VLE-IG

☐ Estimate missing parameters by UNIFAC

Figure 7.9   Alternative data sources.

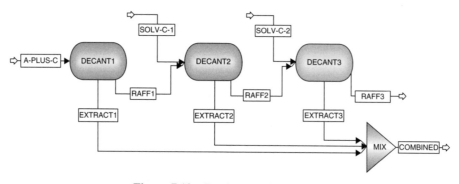

Figure 7.10   Batch extraction process.

time step later in the process, when it contains the remaining raffinate and is charged with fresh solvent C. The composition of the mixed extracts is the product of the mixer.

7.4 WORKSHOPS

Workshop 7.1   Use the Flash2 block to flash the feeds in Table 7.1 adiabatically at constant pressure for the purpose of reducing the acetone composition in the bottoms product. The feed temperature is 70°C and its pressure is 760 mmHg.

Be sure to set up appropriate units. Use Aspen Plus's stored Uniquac, NRTL, and Wilson parameters as the basis for the calculations. What do you observe?

TABLE 7.1   Feeds for Workshop 7.1

Component	Feed 1 (lbmol/hr)	Feed 2 (lbmol/hr)
Ethanol	40	
Water		50
Acetone	10	

Workshop 7.2   Based on the solution to Workshop 7.1, using 760 mmHg as a base case, create a sensitivity study for pressure varying between 10 and 760 mmHg. Show the effect of varying flash pressure on fraction vaporized, mole fraction acetone in the vapor and liquid, and flash temperature using Aspen Plus's stored Uniquac parameters.

Workshop 7.3   The feeds in Table 7.2 are to be fed to a Flash3 block at a pressure of 760 mmHg and a temperature of $80°C$. Using Aspen's stored Uniquac parameters, use a sensitivity study to find the temperature at which the vapor phase disappears when operating at 760 mmHg. The temperature and pressure of both feeds are $25°C$ and 760 mmHg.

TABLE 7.2   Feeds for Workshop 7.3

Component	Feed 1 (lbmol/hr)	Feed 2 (lbmol/hr)
Ethanol	10	
Water		40
Toluene	50	

Workshop 7.4a   Using the feeds from Workshop 7.3, employ the Decant block to calculate the composition and quantity of the resulting phases at $25°C$ using the parameters provided in Table 7.3. Be sure to set up appropriate units.

TABLE 7.3   Uniquac Parameters for a Toluene–Water–Ethanol System

Parameter/ Element No.	Component Pair	Parameter Estimate (SI units)
UNIQ/2	Toluene–H_2O	-926.6623
UNIQ/2	H_2O–Toluene	-258.8106
UNIQ/2	EtOH–H_2O	301.8589
UNIQ/2	H_2O–EtOH	-119.5998
UNIQ/2	EtOH–Toluene	-49.37053
UNIQ/2	Toluene–EtOH	-100.2406

Workshop 7.4b   Develop a sensitivity study to determine the effect of temperature on this operation. Vary the temperature from 30 to $100°C$, displaying the amount of ethanol extracted into the water-rich phase and the ethanol composition of the toluene-rich phase. Be sure that the base case is executed last. Note all observations concerning this sensitivity study. Are the results reasonable?

Workshop Notes

Workshop 7.1 A duplicator block will permit all three simulations to be done at once, and the results will facilitate easy comparison. Two of the results are comparable. The third displays a similar trend, but which are correct?

- Do the activity coefficient models work for this system?
- Were the parameters in the database fit properly?
- Should new data be located and custom-fit?
- Would an "in the field" design be acceptable for controlling the product composition (e.g., a secondary sensor loop feeding the set point of the primary control loop, to make up for model deficiencies)?

Workshop 7.3 A sensitivity study with a plot should be a good guide for a design specification. See the solutions to Workshops 7.4a and 7.4b.

Workshop 7.4 All Uniquac parameters except Uniq/2 equal must be set to zero.

- Where did the list of Uniquac parameters come from? How can their quality be assessed?
- Results show that extraction improves as the temperature gets colder.

REFERENCES

Poling, B. E., Prausnitz, J. M., and O'Connell, J. P., *The Properties of Gases and Liquids*, 5th ed., McGraw-Hill, New York, 2000.

Prausnitz, J. M., Lichtenthaler, R. N., and de Avezedo, E. G., *Molecular Thermodynamics of Fluid-Phase Equilibria*, 3rd ed., Prentice Hall, Upper Saddle River, NJ, 1999, p. 41.

Walas, S. M., *Phase Equilibria in Chemical Engineering*, Butterworth, Woburn, MA, 1985, p. 388.

CHAPTER EIGHT

PRESSURE CHANGERS

Aspen Plus offers six different types of pressure changers. The Pump, Compressor, and Multistage Compressor models actively change pressure as a result of the addition of some form of work to the block which is converted to action on a stream. The three other models—valve, pipe (single segment with fittings), and pipeline (multiple segments with fittings)—are stationary objects that incorporate resistance to flow which must be overcome by use of a pump or compressor.

8.1 PUMP BLOCK

The Pump block typically is used to calculate the power required to pump a fluid up to a specified pressure. The Pump block can be used to model a hydraulic turbine as well. In this variation the block calculates the power produced when the discharge pressure is specified. Alternatively, the power input can be specified and the pump will calculate the discharge pressure. Figure 8.1 shows the Pump and Turbine input options. Details of the calculations are well documented in Aspen Plus's Help under Pump Reference. An example of the use of a pump and a hydraulic turbine may be found at Chapter Eight Examples/pumps. For cases in which meaningful pumping applications are not needed, a Heater block can be used to specify the state of a stream.

8.2 COMPR BLOCK

Details of single-stage compressor modeling can be found in Aspen Plus's Help under the Compr Reference, with a variety of links to all variants. An excellent overview may be found by following the link Mollier method, Compressor Background. These

Teach Yourself the Basics of Aspen Plus™   By Ralph Schefflan
Copyright © 2011 John Wiley & Sons, Inc.

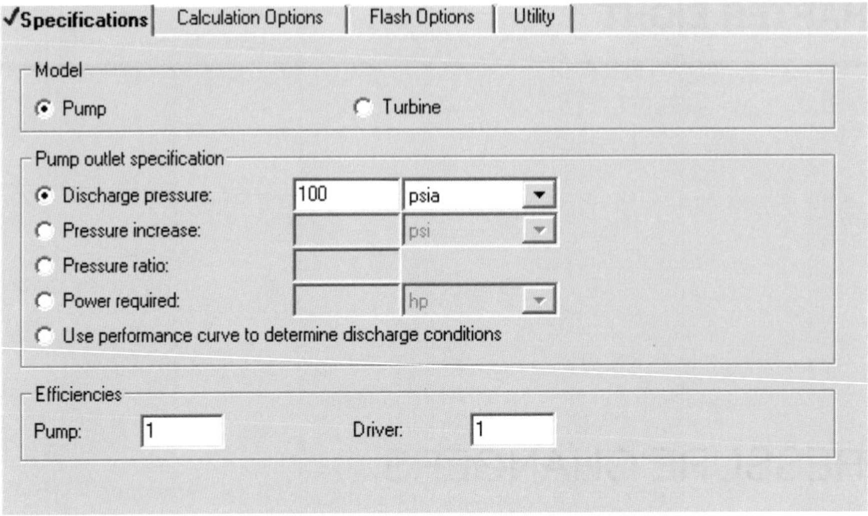

Figure 8.1   Pump and turbine input.

sections should be read prior to using the block. The basic idea is given by

$$\eta \Delta h = \int_{p_1}^{p_2} V \, dp \tag{8.1}$$

where η is either a polytropic or isentropic efficiency factor which is assumed constant for a process, where Δh is the change in enthalpy between the two states, V the volume, and p the pressure. The various methods available in Aspen Plus are based on assumptions dealing with how the integral is evaluated.

The polytropic efficiency is given by

$$\frac{n-1}{n} = \frac{(k-1)/k}{\eta_P} \tag{8.2}$$

where n is the polytropic coefficient, with each state described by pV^n, k is the heat capacity ratio c_p/c_v, and η_p is the polytropic efficiency.

For an isentropic process h_{out} is obtained from the outlet pressure when the entropy at the outlet equals the entropy at the inlet. The isentropic efficiency is used to modify an isentropic process to approach a real condition. There are two forms of isentropic efficiency:

$$\eta_s = \frac{h_{out}^s - h_{in}}{h_{out} - h_{in}} \tag{8.3}$$

for compression and

$$\eta_s = \frac{h_{out} - h_{in}}{h_{out}^s - h_{in}} \tag{8.4}$$

for expansion.

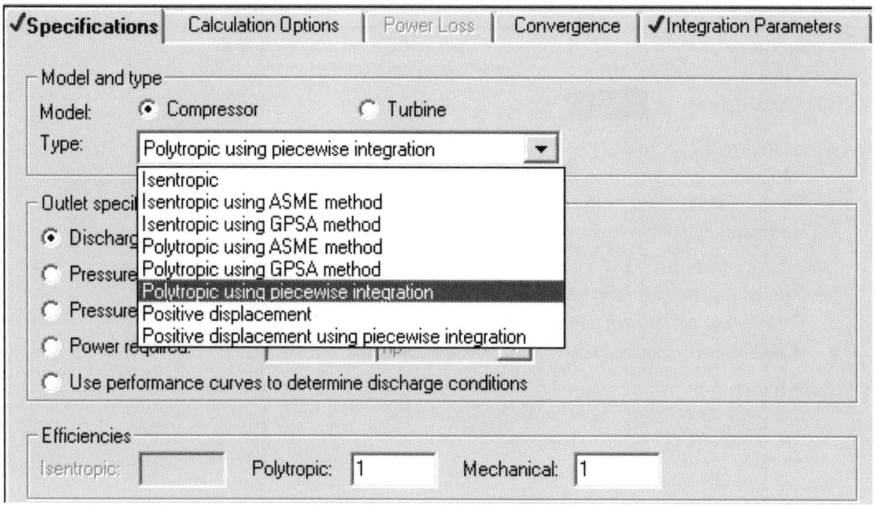

Figure 8.2 Compr options.

It is also necessary to specify the mechanical efficiency, defined by

$$\text{IHP} = F \Delta h \tag{8.5a}$$

$$\text{BHP} = \frac{\text{IHP}}{\eta_m} \tag{8.5b}$$

where F is the molar flow rate, Δh the enthalpy change, IHP the indicated horsepower, BHP the brake horsepower, and η_m the mechanical efficiency.

The Aspen Plus Compr block can be used to model a compressor or turbine. Three models of compressors are available: a polytropic centrifugal, a polytropic positive displacement, and an isentropic compressor, each with variants as shown in Figure 8.2. If a turbine is chosen, isentropic is the only possibility. Some examples of Compr applications are given at Examples/compressor.

8.3 MCompr BLOCK

Details of multistage compressor modeling can be found in Aspen Plus's Help under the MCompr Reference, and the link specifying MCompr. These sections should be read prior to using the block. The MCompr block can be used to model six types of compressors, as shown in Figure 8.3. The tabs Specs and Cooler are used to specify the various efficiencies and the stage outlet conditions, respectively. The Convergence tab is used to select the allowed phases in the compressor and cooler stages. An example of a MCompr application is given at Examples/mcompr.

8.4 PIPELINES AND FITTINGS

The Pipe block calculates the pressure drop in a single run of pipe. Pipe permits the connection of fittings. The flow is assumed to be uniform along the axis and can

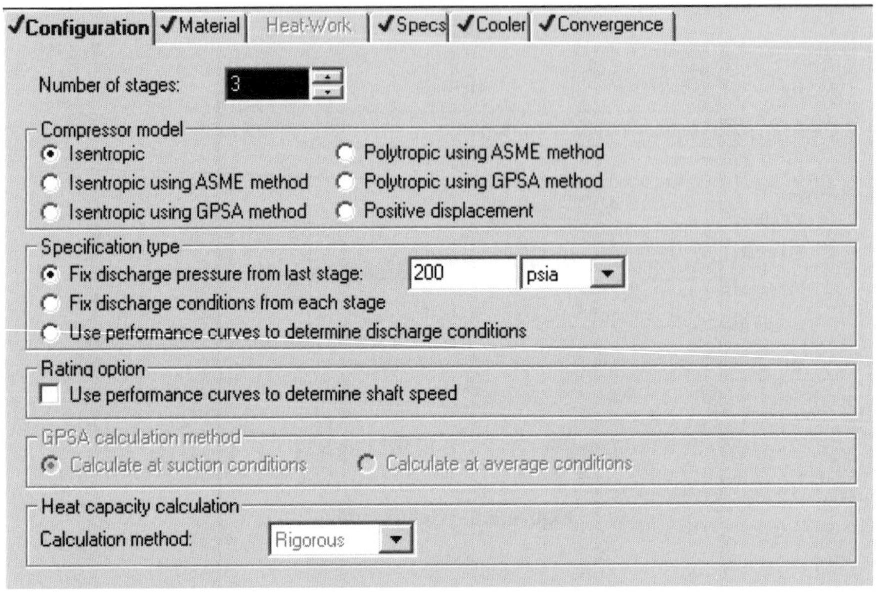

Figure 8.3 MCompr options.

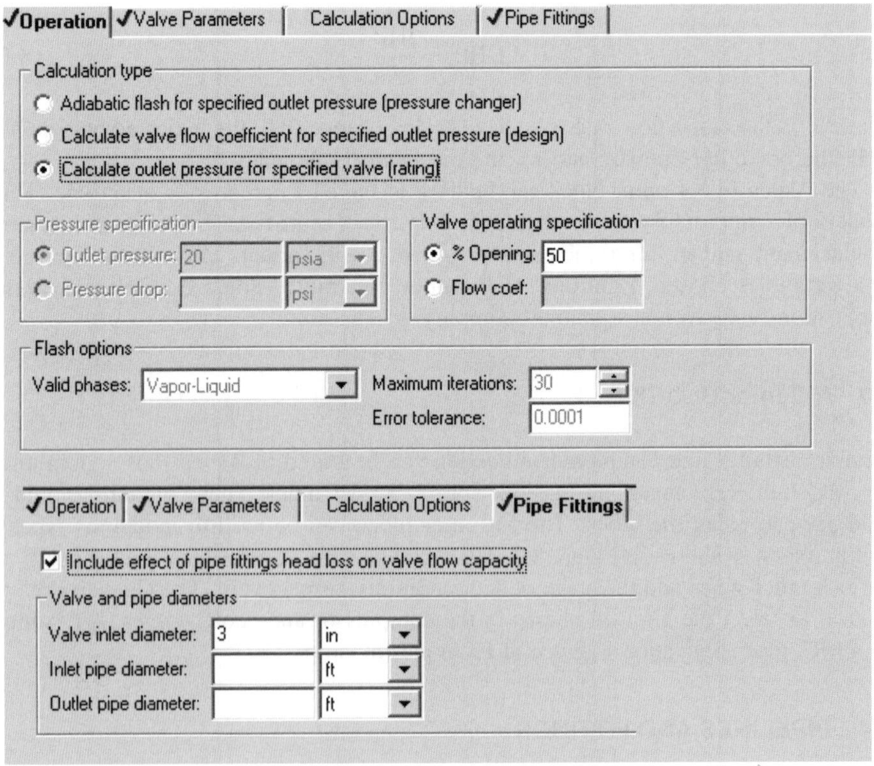

Figure 8.4 Pipeline and fittings.

perform one, two, or three phase calculations. If there are multiple runs of pipe of different diameters or elevations, the Pipeline block should be used. Figure 8.4 shows the Pipe input options. An example of usage is given at Examples/pipeline.

8.5   WORKSHOPS

Workshop 8.1   The stream whose composition is given in Table 8.1 is to be pumped from 25 to 100 psi. The inlet temperature is $-10°C$. Use the Pump block to calculate the horsepower required if the pump's efficiency is 80%.

TABLE 8.1   Feed for All Chapter 8 Workshops

Component	Abbreviation	Lbmol/Hr
Methane	C1	0.1
Ethane	C2	1
Propane	C3	10
n-Butane	NC4	18.9
1-Butane	1C4	20
1,3-Butadiene	DC4	20

Workshop 8.2   The feed of Workshop 8.1 is a vapor at 100 psi and 100°C. It is to be compressed to 500 psi using a polytropic compressor with an efficiency of 100%. The efficiency of the motor driving the compressor is also 100%. Use the Compr block to calculate the performance of the compressor and the outlet state of the product.

Workshop 8.3   Repeat Workshop 8.2 using an isentropic compressor that operates reversibly (i.e., its isentropic efficiency is 100%). The efficiency of the motor driving the compressor is also 100%.

Workshop 8.3a   Repeat Workshop 8.3 with an isentropic efficiency of 0.8.

Workshop 8.4   Repeat Workshop 8.2 using a two-stage isentropic compressor where both stages and motors have 100% efficiencies. No interstage coolers are to be used. All pressure drops are zero. Use the block MCompr to calculate the compressor's performance and the state of the outlet stream. Compare results to those of Workshop 8.1.

Workshop 8.5   Using Workshop 8.4 as a base case, run case studies to determine the effect of an intercooler between stages 1 and 2 of the compressor. For the case studies, specify the heat removed as 0, 50,000, 100,000, 175,000, and 200,000 Btu/hr. Show the effect of heat removal on total brake horsepower, total heat removal, temperature of the gas at the outlet of stage 1 of the compressor, and temperature of the gas at the outlet of the cooler.

Workshop 8.5a   Repeat the 175,000-Btu/hr heat removal case from Workshop 8.5 but include a liquid removal stream.

Workshop Notes

Workshop 8.2 and 8.3   These two workshops permit comparison of polytropic and isentropic compressors.

Workshop 8.3 and 8.3a   These two workshops permit comparison of ideal and real isentropic compressors.

Workshop 8.4   This is a straightforward two-stage MCompr example.

Workshop 8.5 and 8.5a   Heat removal requires a minus sign. Note the warning accompanying the $-17{,}500$-Btu/hr removal case. The effect of an intercooler is shown. The warning disappears in Workshop 8.5a when liquid removal is provided for.

REFERENCE

Aspen Plus version 7.0, Help: Pump Reference, Compr Reference, and MCompr Reference.

CHAPTER NINE

HEAT EXCHANGERS

The implementation of heat exchanger models is different from that of most Aspen Plus models in that some are capable of detail design using very high quality industrial programs which are integrated into Aspen Plus. These are the Hetran and Tasc shell-and-tube heat exchanger and the Aerotran air-cooled exchanger programs. They are documented in Aspen Plus Help: EDR (Exchanger Design and Rating). Most can be used for design as well as being used to rate existing exchangers; that is, they can be used as sequential modular models.

To simulate a heat exchanger one must solve the primary equations

$$q - m_{\text{hot}} c_p^{\text{hot}} (T_{\text{in}}^{\text{hot}} - T_{\text{out}}^{\text{hot}}) = 0 \tag{9.1}$$

$$q - m_{\text{cold}} c_p^{\text{cold}} (T_{\text{out}}^{\text{cold}} - T_{\text{in}}^{\text{cold}}) = 0 \tag{9.2}$$

$$q - UAF \, \Delta T_{LM} = 0 \tag{9.3}$$

$$\Delta T_{LM} = \frac{\Delta T_1 - \Delta T_2}{\ln(\Delta T_1 / \Delta T_2)} \tag{9.4}$$

Here q is the exchanger duty; m a flow rate, c_p the heat capacity; T the temperature; ΔT the temperature difference at an end of the exchanger; U an overall heat transfer coefficient which depends on temperature, transport properties, and exchanger geometry; and F a correction factor for multiple tube-side and/or shell-side passes. The factor F, derived through the work of Nagle (1933) and Underwood (1934), can be calculated as

$$F = \frac{\sqrt{R^2 + 1}\ln[(1 - S)/(1 - RS)]}{(R - 1)\ln\frac{2 - S(R + 1 - \sqrt{R^2 + 1})}{2 - S(R + 1 + \sqrt{R^2 + 1})}} \tag{9.5}$$

where

$$R = \frac{T_{in}^{hot} - T_{out}^{hot}}{T_{out}^{cold} - T_{in}^{cold}} \tag{9.6}$$

$$S = \frac{T_{out}^{cold} - T_{in}^{cold}}{T_{in}^{hot} - T_{in}^{cold}} \tag{9.7}$$

When used in simulation mode, the state of the exchanger feeds must be specified. Depending on the complexity of the model chosen, the heat transfer area, A, is either specified or calculated from the heat exchanger physical layout. Heat transfer coefficients are calculated from correlations such as the Hewitt (1992) correlation prepared by Gnielinski, which involves the Nusselt $N_{NU} = hD/k$, Reynolds $N_{RE} = DG/\mu$, and Prandtl $N_{PR} = c_p\mu/k$ numbers, and the Darcy friction factor, given by

$$N_{NU} = \frac{h_i D_i}{k} = \frac{(f_D/8)(N_{RE} - 1000)N_{PR}}{1 + 12.7\sqrt{F_D/8}(N_{PR}^{2/3} - 1)}\left(1 + \frac{D^{2/3}}{L}\right) \tag{9.8}$$

$$f_D = (1.82\log_{10}N_{RE} - 1.64)^{-2} \tag{9.9}$$

Here h_i is the inside pipe heat transfer coefficient, D_i the inside pipe diameter, k the thermal conductivity, G the mass flow rate, μ the viscosity, c_p the heat capacity, and f_D the Darcy friction factor. Complete documentation of all correlations used in Aspen Plus can be found in Help: Heatx Reference and Model Reference. Depending on which model is chosen, U is either specified or calculated iteratively during the convergence process. The four heat-exchange-related models can be found in the model library under the tab Heat Exchangers.

9.1 HEATER BLOCK

An example of the primary input form of the Heater block, which shows the possible specifications, is shown in Figure 9.1. The Heater block offers a variety of ways to specify the output stream state, all of which result in calculation of the energy required to heat (or cool) a stream. Alternatively, one may specify the energy added to or removed from a heater, which is used by the block to establish the state of the output stream.

An important capability is the use of two heaters to model a heat exchanger bypassing the use of equations (9.3) and (9.4), as shown in Figure 9.2. Note the use of a heat stream to connect the two heaters. The heat stream should be aligned in the correct direction, which depends on which heater will receive the heat, either positive or negative. Care must be taken with the sign of the heat transferred (heat added is positive). In this example the outlet temperature of heater H2 is specified and the heat, stream 5, flows to heater H1. This example may be found at Chapter Nine Examples/Heaters.

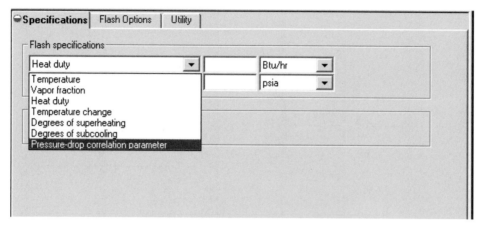

Figure 9.1   Heater block primary input.

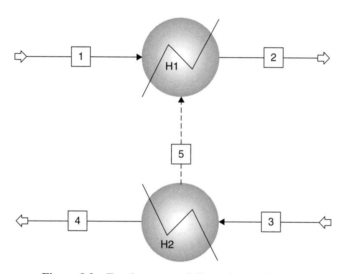

Figure 9.2   Two heaters modeling a heat exchanger.

An example of a process in which heat exchangers are replaced with heaters is shown in Figure 9.3. The heat exchangers here are used to conserve energy by extracting heat from the bottoms of the columns and transferring part of the available energy to the column feeds. An important aspect of using heaters rather than heat exchangers is that the calculations can be performed once through, and no tear streams are required. This is illustrated by Aspen Plus's analysis of the process, given in Figure 9.4. In most cases, whenever a heat exchanger is placed on a flowsheet, a tear stream is introduced. This is illustrated in Figure 9.5, where the heaters from the Figure 9.3 process are replaced by heat exchangers. Aspen Plus's analysis, in Figure 9.6, shows that one tear stream has been generated for each heat exchanger inserted. Details of the solution to the process with four heaters are given at Examples/fourheatersflowsheet. Similarly, the two-heat-exchanger process is provided at Examples/twoheatxflowsheets.

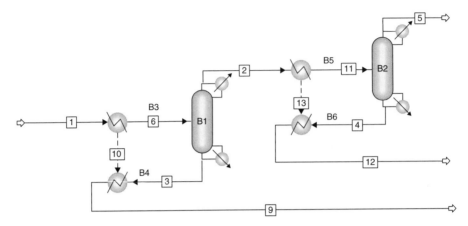

Figure 9.3   Process with two heaters as heat exchangers.

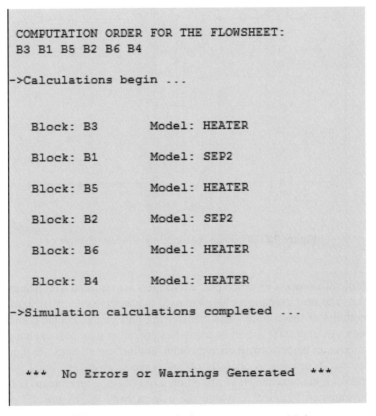

Figure 9.4   Computation sequence for a process with heaters.

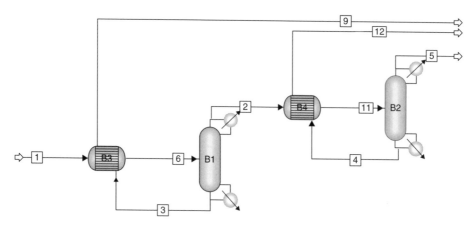

Figure 9.5 Process with two heat exchangers.

```
 Flowsheet Analysis :

Block $OLVER01 (Method: WEGSTEIN) has been defined to converge
      streams: 3

Block $OLVER02 (Method: WEGSTEIN) has been defined to converge
      streams: 4

COMPUTATION ORDER FOR THE FLOWSHEET:
$OLVER01 B3 B1
(RETURN $OLVER01)
$OLVER02 B4 B2
(RETURN $OLVER02)
```

Figure 9.6 Tear streams for a process with heat exchangers.

9.2 HEATX BLOCK

The Heatx block has very wide applicability, as shown in its primary input form shown
in Figure 9.7. To use a Heatx exchanger in a simulation, it must have been designed so
that it can be used as part of the sequential modular approach. Note that Heatx permits
a design, a rating, and a simulation mode. The simulation mode requires that the inputs
to the exchanger be either specified or calculated from a previous block, because the
block is expected to behave as a sequential modular model. The rating mode permits
the specification of a block output as shown in Figure 9.8 and behaves as if it were a
sequential modular model with a design specification imposed on an output.

The simplest approach for using the block for design is to select a shortcut design,
in which case an output of an exchanger is specified. The available specifications for
design are shown in Figure 9.7. The solution to the design problem, using the shortcut
method, yields a value for the *UA* product which can be employed as part of the
specifications when setting up a simulation problem. An example of this approach
is given in the solution to the flowsheet shown in Figure 9.5, which incorporates

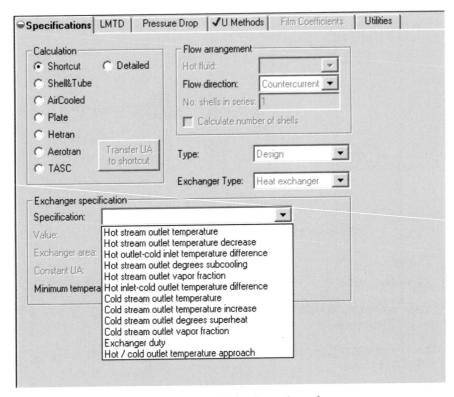

Figure 9.7   Heatx block primary input form.

exchanger design. The specification used for design, the outlet temperature of the cold stream, which was used for both exchangers, was transferred from the four-heater flowsheet to the two-exchanger flowsheet. The *UA* values calculated for the shortcut design of two exchangers were transferred to the simulation flowsheet to produce the completed process simulation. Details may be found at Examples by comparing twoheatxflowsheetd (design) and twoheatxflowsheets (simulation). When using rating or simulation calculations, both the shortcut and detailed options permit the user to enter estimates of the overall heat transfer coefficients from sources such as Perry and Green (1999). In addition to an overall heat transfer coefficient, the heat transfer area must be provided.

The detailed option permits only rating and simulation modes, both of which can be used as part of a sequential modular simulation. In both cases the selection of geometry prompts a user for the detailed layout of a shell-and-tube heat exchanger. A detailed design must be completed prior to the selection of this option.

As can be seen from the calculation frame in Figure 9.7, Aspen Plus provides a rich selection of detailed design calculation methods. These are referred to as Aspen Exchanger Design and Rating. An example of the use of "shell and tube" is given by redesigning the first exchanger at Examples/twoheatxflowsheetd. The input is virtually identical to that at Examples/twoheatxflowsheetd except for the identification of the hot fluid and the specification of an input file, as shown in Figure 9.9. This example is given at Examples/B3shellandtube. The file B3shellandtube.txt gives a report of the

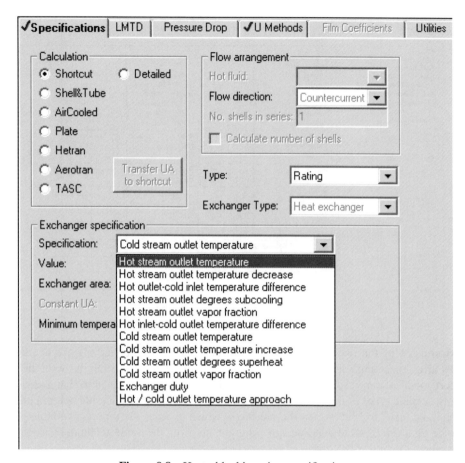

Figure 9.8   Heatx block's rating specifications.

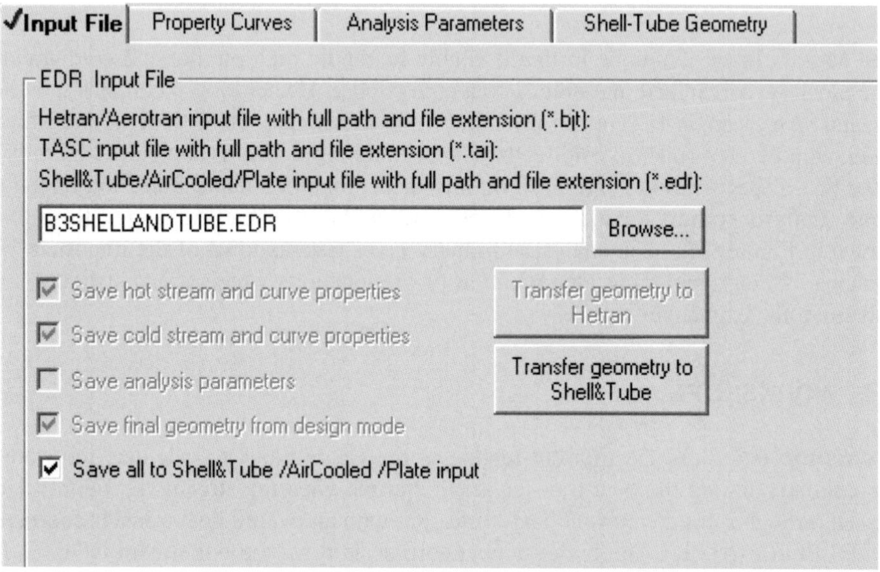

Figure 9.9   Mheatx primary input form.

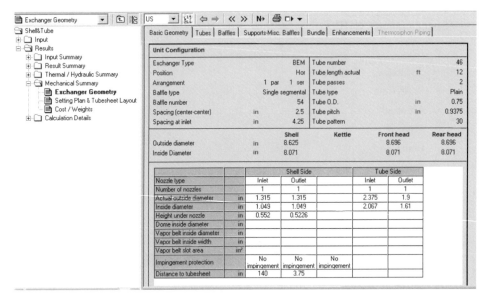

Figure 9.10 Mheatx zone output.

performance of the exchanger; however, mechanical and other details are available only after executing the B3shellandtube.bkp file and using the EDR browser in the report mode of the results. An example is given in Figure 9.10. Note that there are many detailed types of outputs which are selected from the left-hand-side column under "shell and tube" by clicking on the plus sign, which exposes the reports available. The choice of a different detailed design program involves the same techniques as those described above.

9.3 MHEATX BLOCK

The Mheatx block Is unique in that it is able to handle multiple hot and cold streams. The block is reorganized internally into a collection of Heater blocks connected by heat streams. An example is given in Examples/mheatxexample. Here there is one Mheatx block with two hot and two cold feeds. The primary input form is shown in Figure 9.11. Note that a specification is required on all but one output stream. Selection of the tab Zone Analysis permits the assignment of zones within the heat exchanger. Results are shown in Figure 9.12. The analysis summary gives a breakdown of the important heat exchange factors, including identification of the locations where streams should enter and leave the exchanger.

9.4 WORKSHOPS

Workshop 9.1 Use the shortcut feature of the Heatx block to calculate the state of the cold stream and the heat transfer area required when the streams in Table 9.1 are passed through a countercurrent heat exchanger with an overall heat transfer coefficient of 100 Btu/(hr-ft^2-°R). The outlet temperature desired for the hot stream is 70°F. The

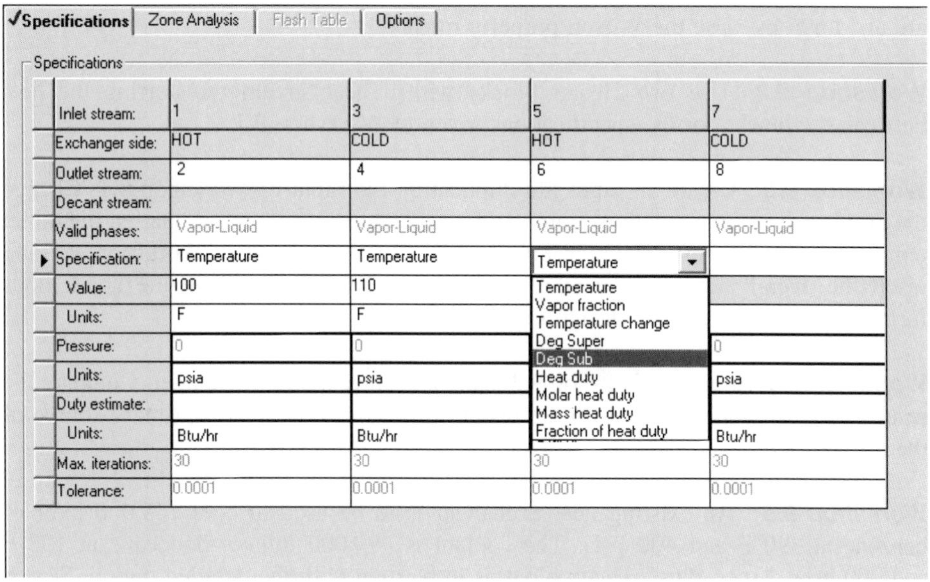

Figure 9.11 Shell-and-tube EDR input file specification.

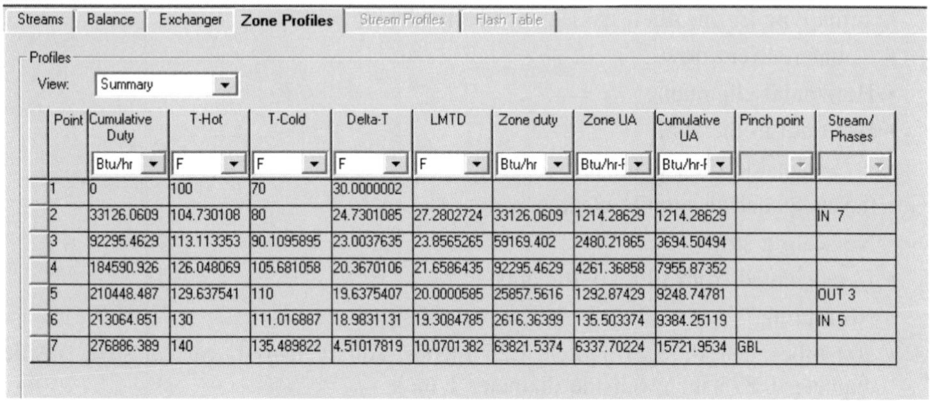

Point	Cumulative Duty	T-Hot	T-Cold	Delta-T	LMTD	Zone duty	Zone UA	Cumulative UA	Pinch point	Stream/ Phases
	Btu/hr	F	F	F	F	Btu/hr	Btu/hr-F	Btu/hr-F		
1	0	100	70	30.0000002						
2	33126.0609	104.730108	80	24.7301085	27.2802724	33126.0609	1214.28629	1214.28629		IN 7
3	92295.4629	113.113353	90.1095895	23.0037635	23.8565265	59169.402	2480.21865	3694.50494		
4	184590.926	126.048069	105.681058	20.3670106	21.6586435	92295.4629	4261.36858	7955.87352		
5	210448.487	129.637541	110	19.6375407	20.0000585	25857.5616	1292.87429	9248.74781		OUT 3
6	213064.851	130	111.016887	18.9831131	19.3084785	2616.36399	135.503374	9384.25119		IN 5
7	276886.389	140	135.489822	4.51017819	10.0701382	63821.5374	6337.70224	15721.9534	GBL	

Figure 9.12 EDR browser in report mode.

TABLE 9.1 Feed for Workshop 9.1

Component	Hot Feed (lb/hr)	Cold Feed (lb/hr)
Methanol	200	
Water	1800	
Ethylene glycol		400

temperatures of the hot and cold feeds are 140 and 40°F, respectively. Both streams are at 14.696 psi. Use the Wilson property method.

Workshop 9.2 Use two Heater blocks with a heat stream transferring the heat between the blocks for the specifications given in Workshop 9.1.

Workshop 9.3 Create an input file duplicating the input of Workshop 9.1. Change the problem type to simulation. Remove the exchanger specification, that is, the outlet temperature of the hot stream. Insert the *UA* value calculated in Workshop 9.1 and insert the overall heat transfer coefficient of 100 Btu/(hr-ft^2-°R) in the U Methods form. Run and compare the design to the simulation.

Workshop 9.4 Repeat Workshop 9.3 using a cocurrent heat exchanger. Compare the results to Workshop 9.3. How do the results compare? What is the primary source of the differences?

Workshop 9.5 An existing heat exchanger is to be used to cool 145,000 lb/hr of benzene at 390°F and 400 psia. The coolant is 490,000 lb/hr *n*-dodecane at 100°F and 200 psia. Aspen Plus's Heatx block is to be used with the detailed option. Search Aspen Plus's Help for exchanger configuration for general information on exchanger geometry.

Details of the exchanger design are as follows:

- Tema type E: one shell pass, two tube passes
- Countercurrent flow
- Horizontal alignment
- Hot fluid in the shell
- 2.75-ft inside shell diameter
- 0.5-inch shell-to-bundle clearance
- No sealing strips
- *U* calculated from film coefficients
- No fouling
- 200 tubes, length 32 ft, pitch 1.25 inches, square layout, carbon steel, inside diameter 0.875 inch; outside diameter 1 inch
- 24 segmental baffles; baffle cut 0.2
- Tubesheet-to-baffle spacing and baffle-to-baffle spacing 1.2 inches
- Tubes in baffle window
- Shell-side nozzles 6 inches
- Tube-side nozzles 8 inches

Use Aspen Plus's defaults when necessary.

Workshop 9.6 Repeat Workshop 9.5 using the Rating Option.

Workshop 9.7 Repeat Workshop 9.5 using the Simulation Option.

Workshop Notes

Workshop 9.1 This is a design problem using shortcut methods. After providing an exchanger specification (i.e., hot stream outlet temperature and an overall heat transfer coefficient, both defined in the workshop description), the design is straightforward.

Workshop 9.2 The results of Workshop 9.1 are duplicated.

Workshop 9.3 As this is a simulation workshop it is necessary to change the specification of Workshop 9.1 to simulation and to insert the value of *UA* calculated by Workshop 9.1. The hot stream specification must also be removed. The simulation duplicates the results of Workshop 9.1.

Workshop 9.4 Use of the exchanger design from Workshop 9.1 but with cocurrent flows results in poorer performance. The heat transferred in Workshop 9.3 is about 131,000 Btu/hr, compared to 99,000 Btu/hr in Workshop 9.4. The log mean temperature difference for the two cases is 35.1 and 26.6°F, respectively, which explains the poorer performance of Workshop 9.4.

Workshop 9.5 The solution is straightforward.

Workshop 9.6 The hot stream specification is taken from the solution of Workshop 9.5. The results are virtually identical to those of Workshop 9.5.

REFERENCES

Aspen Plus version 7.0, Heaters documentation.

Aspen Plus version 7.0, Help: EDR, Exchanger Configuration, Heatx Reference, and Model Reference.

Hewitt, G. F., Ed., *Handbook of Heat Exchanger Design*, Begell House, New York, 1992.

Nagle, W. M., Mean temperature differences in multipass heat exchangers, *Ind. Eng. Chem.*, 25, 604–609 (1933).

Perry, R. H., and Green, D. W., *Perry's Chemical Engineers' Handbook on CD-ROM,* 7th ed., McGraw-Hill, New York, 1999, Table 11.2.

Underwood, A. J. V., The calculation of the mean temperature differences in multipass heat exchangers, *J. Inst. Pet. Technol.*, 20, 145–158 (1934).

CHAPTER TEN

REACTORS

Aspen Plus provides seven different models of reactor blocks that deal with chemical reactors in different ways:

1. *RStoic:* a stoichiometry-based reactor with specified extents of reaction.
2. *RYield:* a reactor based on specified yields.
3. *REquil:* a rigorous equilibrium reactor based on reaction stoichiometry.
4. *RGibbs:* a rigorous reactor which includes phase equilibrium using Gibbs free energy minimization.
5. *RCSTR:* a rigorous stirred-tank reactor with rate-controlled reactions based on specified kinetics.
6. *RPlug:* a rigorous plug flow reactor with rate-controlled reactions based on specified kinetics.
7. *RBatch:* a rigorous batch and semibatch reactor with rate-controlled reactions based on specified kinetics.

The reactor blocks can be found in the model library under the tab Reactors.

10.1   RStoic BLOCK

The RStoic block is the simplest of Aspen Plus's reactor blocks. It permits the use of several reactions with the molar extent of conversion or fractional conversion of a component, specified for each reaction. For the reaction

$$aA + bB \rightarrow cC + dD \tag{10.1}$$

Teach Yourself the Basics of Aspen Plus™   By Ralph Schefflan
Copyright © 2011 John Wiley & Sons, Inc.

the molar extent of reaction X is defined by

$$X = -\frac{\Delta n_a}{a} = -\frac{\Delta n_b}{b} = \frac{\Delta n_c}{c} = \frac{\Delta n_d}{d} \tag{10.2}$$

where Δn_i are the moles of component i either created or consumed. The applicable material balance equations with the feeds and products of component i, defined as F_i and P_i, respectively, are for a reactant. For example, if i is component A,

$$P_i = F_i - aX \tag{10.3}$$

and for a reaction product, for example, if i is component D,

$$P_i = F_i + dX \tag{10.4}$$

This is illustrated at Chapter Ten Examples/RStoicExample and is shown in Figure 10.1. Here a duplicator block feeds each RStoic reactor block. The reaction involved is the formation of ammonia from hydrogen and nitrogen. The reactor ST1 is configured for an extent of reaction of 90 lbmol/hr and reactor ST2 with a fraction converted of 0.50 nitrogen. The details of the extent of reaction configuration is shown in Figure 10.2.

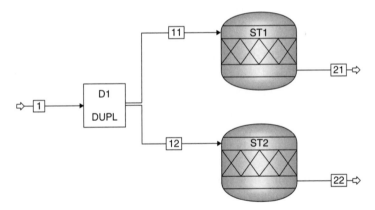

Figure 10.1   RStoic example.

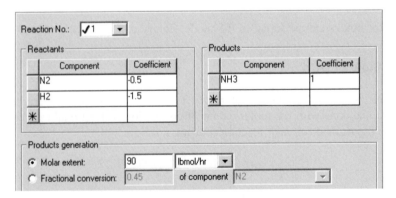

Figure 10.2   Configuration of extent of reaction.

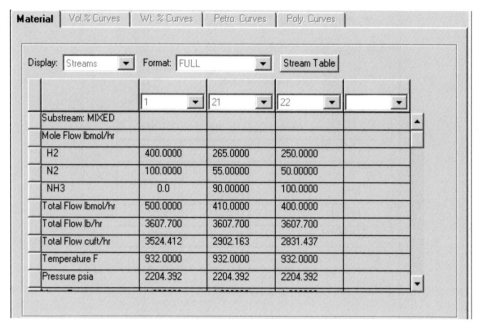

Figure 10.3 Comparison of extent and yield results.

Note that the shaded fraction conversion is 0.45 of N_2, which is equivalent to 90 lbmol/hr of conversion. Aspen Plus does not provide this value. It was placed here for the convenience of the reader. The results of the simulation are given in Figure 10.3, where it may be observed that the fraction converted, 0.5 in ST2, results in 5 additional lbmol/hr of N_2 converted relative to ST1. A value of 0.45 would produce results identical to those of ST1.

10.2 RYield BLOCK

The RYield block is used when the distribution of products is known. No reaction stoichiometry is involved. Two primary options are available: specification of component yields and specification of component mapping. The other two options, a user subroutine and petro specifications, are not considered here.

For the specification of component yields, the yield of each reaction product is specified as either moles or mass per unit mass of feed. A component may be specified as an inert. The yields will then be based on a unit mass of noninert feed. Calculated yields are normalized to maintain an overall material balance. Because of this, yield specifications are actually expressed as ratios of components in the product of the reactor. Yield specifications result in a yield distribution rather than absolute yields. An example of the use of the component yield option may be found at Examples/RYieldExample1. This example involves the reaction of benzene and chlorine to form mono- and dibenzyl chloride with hydrochloric acid as a by-product. In the absence of kinetic data the ratio of products is specified as yields. The yields for this example are shown in Figure 10.4. It should be noted that the multiplication of the yield data by a factor of 10

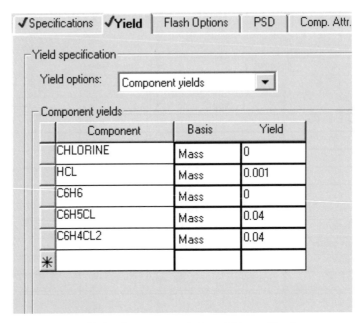

Figure 10.4   Component relative yield specifications.

produces the same results as Example1, which demonstrates that the yield distribution is independent of the numerical values of the yields, but merely on the component ratios. This is easily verified by temporarily changing the yield data in RYieldExample1, executing, and comparing results to those in the file RyieldExample1.txt, which is based on the original specifications.

The same set of reactions is used to demonstrate the component mapping option of the RYield block. No reaction stoichiometry is employed. A reaction can be characterized as lumped, in which case a list of reactants are lumped together to form a specified product. Each of the reactants is assigned a coefficient which defines the weight fraction of the reactant converted to the product.

When a reaction is characterized as delumped, the reactant selected produces a specified list of product components, and similar to lumping, each product is assigned a coefficient which represents the weight fraction of the reactant that is converted to each product. An example of the use of the component yield option may be found at Examples/RYieldExample2. Figure 10.5 shows the implementation of component mapping for the benzene chlorination example. Selecting a reactant and choosing Edit permits selection of the lumping or delumping coefficients.

In both block options selection of inconsistent yield factors or component mapping factors may produce material balance discrepancies. In this case, Aspen Plus normalizes the results to maintain the material balance while maintaining the component ratios calculated.

10.3   REquil BLOCK

The REquil block is used when the reaction stoichiometry is known and some of the reactions reach chemical equilibrium. Additionally, the block imposes phase

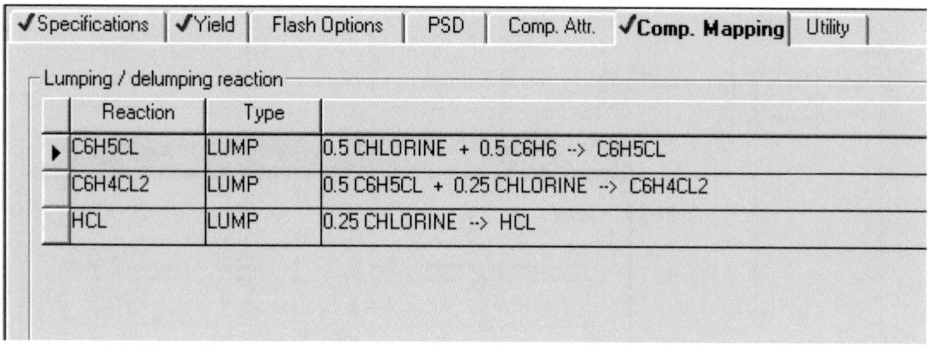

Figure 10.5 Lumping and component mapping.

equilibrium on the products of reaction. Two primary options are available for each reaction: specification of the extent of reaction as described by equations (10.1) through (10.4), or calculation of the equilibrium constant for the reaction K_{eq} from the standard state change Gibbs free energy, $\Delta G°$, of the components of the reaction, using the equation

$$K_{eq} = e^{-(\Delta g°/RT)} \tag{10.5}$$

where

$$\Delta g° = \Sigma\, g°_{\text{products}} - \Sigma\, g°_{\text{reactants}} \tag{10.6}$$

The effect of temperature on the reaction equilibrium constant can be calculated as

$$\ln \frac{K_{eq}^{T_2}}{K_{eq}^{T_1}} = \frac{1}{R} \int \frac{\Delta h°}{T^2}\, dT \tag{10.7}$$

Subsequently, the compositions of the products and reactants at equilibrium are calculated using equation (10.7):

$$K_{eq} = \frac{\Pi\, f_{\text{product}}}{\Pi\, f_{\text{reactants}}} \tag{10.8}$$

where f is the fugacity of a component.

An example of REquil, the chlorination of benzene, using the approach to the equilibrium option may be found at Examples/REquilExample1. The reaction setup and specifications are shown in Figure 10.6. Note that the selection "temperature approach" is $0°F$, which indicates equilibrium. An example of REquil using the "extent of reaction" option, with the same reactions, may be found at Examples/REquilExample2.

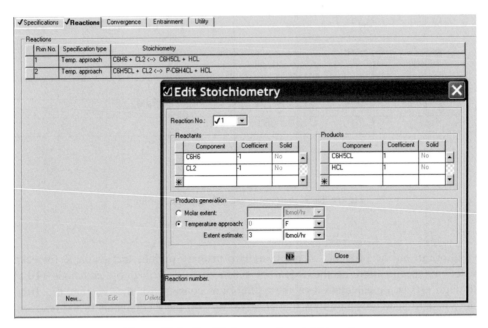

Figure 10.6 Equilibrium reaction stoichiometry.

10.4 RGibbs BLOCK

The RGibbs block can be used to establish the chemical equilibrium composition between reactants and products. Consider the reaction of isomer A to isomer B at a given temperature and pressure. For an n-component mixture where x_i is the mole fraction of component i, the Gibbs free energy g_{mix} is given by

$$g_{mix} = \Sigma\, x_i + RT\, \Sigma\, x_i \ln x_i \qquad (10.9)$$

Differentiating equation (10.9) with respect to x_a and recognizing that for a two-component mixture $x_b = 1 - x_a$, one obtains

$$\frac{dg_{mix}}{dx_a} = g_a^\circ - g_b^\circ + RT \ln \frac{x_a}{1 - x_a} \qquad (10.10)$$

The minimum of g_{mix} can be found by setting equation (10.10) to zero and solving for x_a. The same approach can be taken for complex chemical and phase equilibria. This method is especially useful for liquid–liquid equilibrium, as it does not produce spurious solutions as may be the case for models such as Decant and Flash3, which solve the describing equations with the appropriate activity coefficient equations.

An example of the use of the RGibbs block for the chlorination of benzene is given at Examples/RGibbsExample1. Figure 10.7 shows the basic setup and calculation options available. Figure 10.8 shows detailed specifications and product options. It is important to note that no reactions need to be defined. One may optionally choose to eliminate from consideration components from the mixture. Examples/RgibbsExample2 is used to illustrate the possibility of excluding a possible product. In this case the component

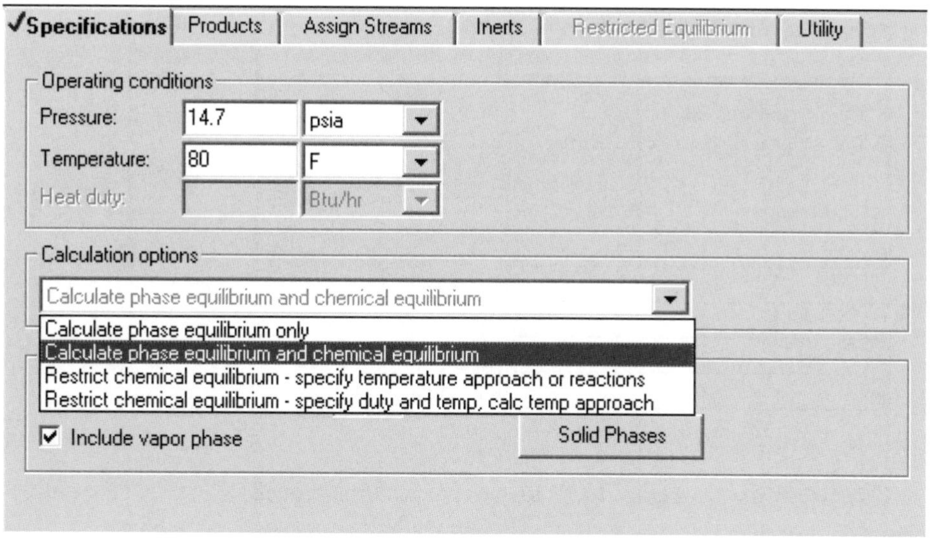

Figure 10.7 RGibbs basic specifications.

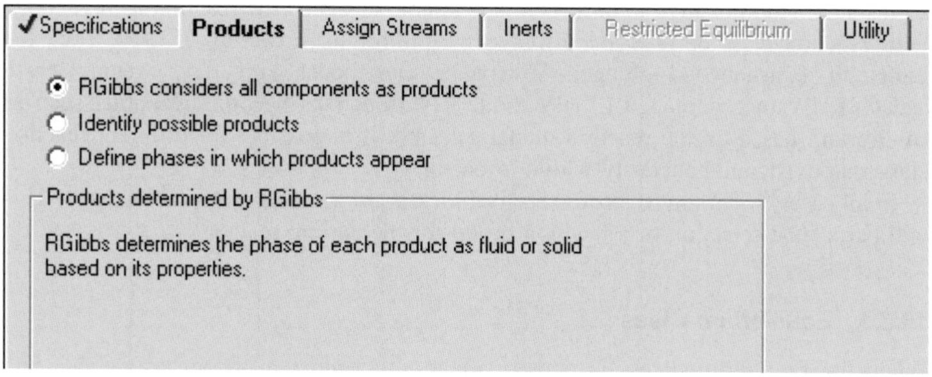

Figure 10.8 RGibbs product options.

chlorobenzene has been removed, as shown in Figure 10.9, and indeed, a look at the solution shows that the chlorination benzene to chlorobenzene does not take place. For this system of reactions, the block offers the possibility to consider additional products: for example, the inclusion of other dichloro isomers in the list of components.

10.5 REACTIONS FOR THE RIGOROUS MODELS

Prior to using the rigorous models, it is necessary to completely define the reactions to be considered. The Reactions setup may be found at the Data Browser display. When the Reactions folder is opened, two possibilities are presented, Chemistry for electrolytes, and Reactions for all others. When reaction(s) are configured they are assigned an identifier, such as R-1, for use in a reactor. As an example,

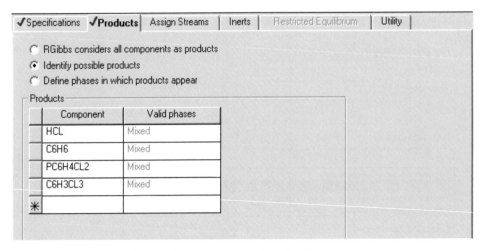

Figure 10.9 RGibbs product identification.

Figure 10.10 shows the configuration of one of the reactions that occurs during the chlorination of benzene. Note that reactant coefficients require a minus sign and products a plus sign. Additionally, a check box is available to define the reaction as reversible. Four reaction class options are displayed: Equilibrium, Powerlaw, LLHW [i.e., the Langmuir–Hinshelwood–Hougen–Watson kinetics model (Perry and Green, 1999)], and GLLHW (a generalized LLHW model). Various tabs become available, such as equilibrium based on the reaction model selected. It is possible for different reaction classes to exist simultaneously within a reactor.

In all cases, when any rigorous reactor block is used, a Reactions tab appears. This will permit the selection of a reaction group for the current reactor.

10.5.1 Equilibrium Class

When the Equilibrium class is chosen, the Equilibrium tab becomes available and Figure 10.11 is displayed. The choice of equilibrium constant by Gibbs energies or by a built-in empirical expression is available, as is the choice of the phase in which the reaction takes place.

10.5.2 Powerlaw Class

When the Powerlaw class is chosen, the Kinetic tab becomes available and Figure 10.12 is displayed. It is important to note that there are six different choices for the concentration of a component. For example, when molarity is chosen from the drop-down list, the appropriate unit, kmol/cubic meter, is displayed but is hidden temporarily underneath the drop-down list. The Powerlaw kinetic expression is composed of two parts, the kinetic factor and the driving force. For all cases the kinetic factor is given by

$$r = k \left(\frac{T}{T_0} \right)^n e^{(-E/R)(1/T - 1/T_0)} \tag{10.11}$$

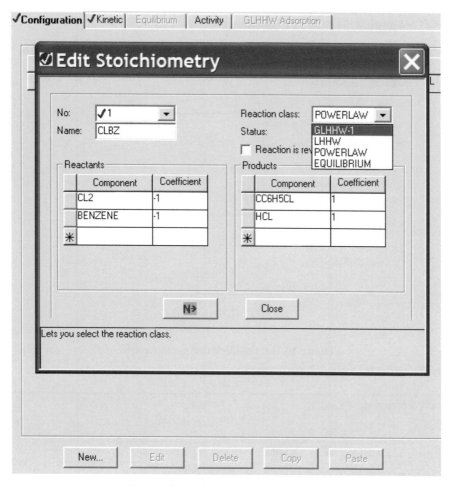

Figure 10.10 Powerlaw kinetic options.

where k is the reaction rate constant, T and T_0 are the reaction and reference temperatures, E, is the activation energy, and n, is an arbitrary coefficient. The input for the kinetic factor can be configured without a reference temperature by setting n and T_0 equal to zero. Setting the activation energy to zero leaves only the reaction rate constant. The driving force for the reaction depends on the concentration basis selected. For example, when the Driving Force button is pushed and molarity is chosen as the concentration basis, the driving force is the continuous product of the component concentrations raised to exponents defined in Figure 10.13. The rate expression becomes

$$r = k \; \Pi \; C_i^{\alpha_i} \tag{10.12}$$

Searching Aspen Plus's Help for kinetic factors will yield an entry Rate Controlled Reactions, which provides much detailed documentation—most important, the power–law expressions used for each possible concentration basis.

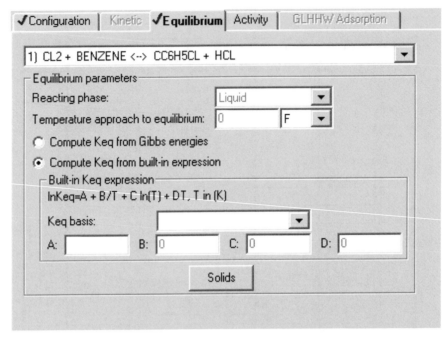

Figure 10.11   Equilibrium constant options.

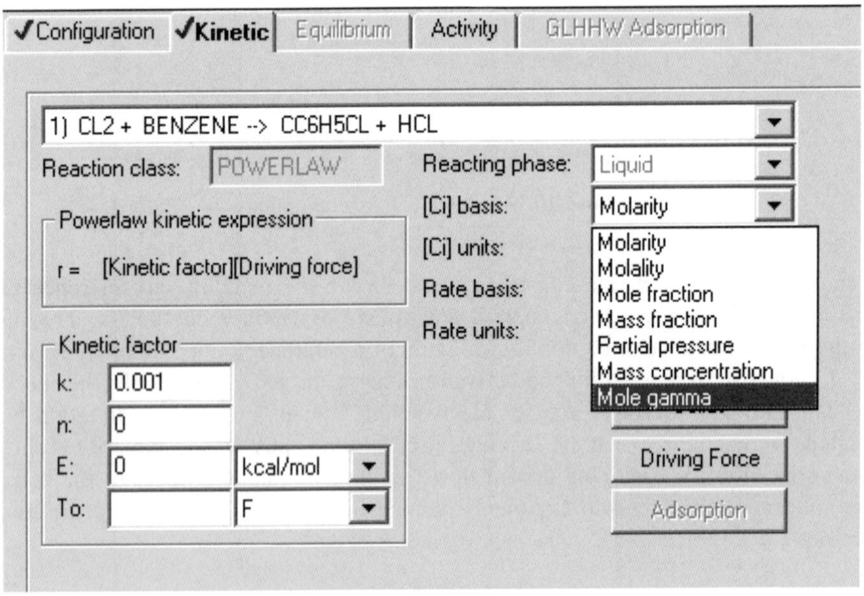

Figure 10.12   Reacting phase options.

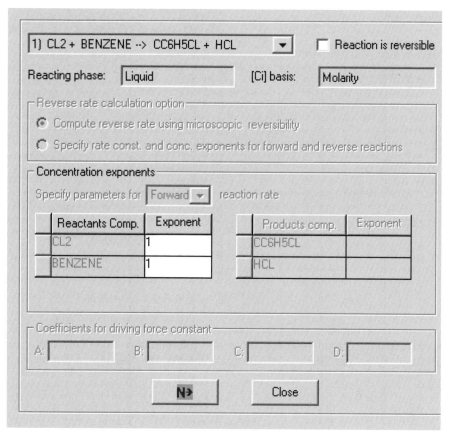

Figure 10.13   Reaction concentration coefficients.

An example of the use of the Reactions entry on the Data Browser with the RCSTR block may be found at Examples/RcstrExample1. This example involves a simplification of the chlorination of benzene to chlorobenzene. The reaction setup for this example is shown in Figures 10.12 and 10.13.

10.5.3   Langmuir–Hinshelwood–Hougen–Watson Class

The LHHW model (Perry and Green, 1999) involves reactions that include adsorption terms in the reaction rate expression

$$\text{rate} = \frac{(\text{kinetic factor})(\text{driving force})}{\text{adsorption term}} \tag{10.13}$$

The kinetic and driving force terms are identical to those in the Powerlaw class. The adsorption expression is given by

$$\text{adsorption expression} = \left[\Sigma \ K_i (\Pi \ C_i^v)\right]_j^m \tag{10.14}$$

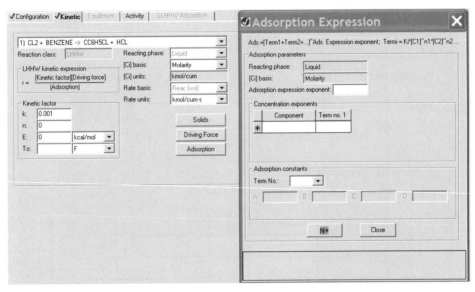

Figure 10.14   LHHW kinetic and adsorption expressions.

where i and j are component indices, v is a concentration exponent, and m is an adsorption expression. The K_i are adsorption equilibrium constants, given by

$$\ln K_i = A_i + \frac{B_i}{T} + C_i \ln T + D_i T \tag{10.15}$$

where T is temperature in kelvin and A, B, C, and D are user-provided parameters.

An example of the reaction setup is given in Figure 10.14. When the adsorption button is pushed, the right-hand display appears and facilitates the entry of adsorption data.

10.5.4   Generalized–Langmuir–Hinshelwood–Hougen–Watson Class

The GLHHW class is designed to enable the user to prepare a customized adsorption term, which is best illustrated by Figure 10.15, which is displayed when the GLHHW tab is selected. Note the equation near the bottom of the figure which begins with Ads.=. This is the complete denominator: that is, the adsorption term in the rate equation. The equation is composed of terms each containing a product of K_i [equation (10.13)] and exponentiated component concentration terms. The exact details of the resulting combination can be seen from the artificial example shown in Figure 10.15. Note that the complete series of products is raised to the power of the adsorption exponent.

10.6   RCSTR BLOCK

The basic ideas describing the analysis of steady-state continuously stirred reactors is well described by Levenspiel (1962). A key element of the analysis is that the

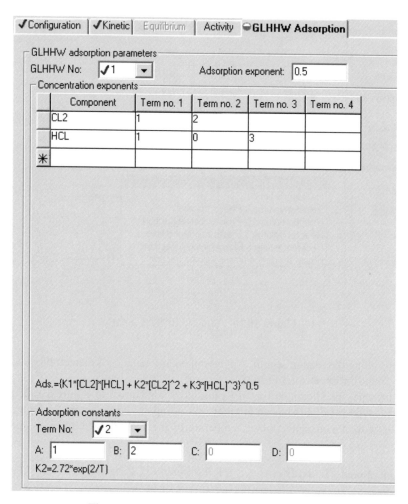

Figure 10.15 GLLHW adsorption parameters.

composition of the reactor is at all times equal to the composition of the reactor product. In essence, for each component entering and leaving a CSTR, the product is calculated by the material balance by accounting for disappearance of the reactant, and accumulation of a product, by the rate of reaction multiplied by reactor volume adjusted using appropriate stoichiometric coefficients. Similarly, an overall energy balance is employed which takes into consideration the energy change due to the reaction.

The primary RCSTR input shown in Figure 10.16 gives all the reactor options available. The minimum additional data required represent the association of the reactor with the reactions involved by using the Reactions entry on the Data Browser. The complete example setup and solution are available at Examples/RCSTRExample1.

10.7 RPlug BLOCK

Unlike the RCSTR block, the composition of an RPlug reactor changes along its length. The basic describing equations involve the integration of appropriate composition and

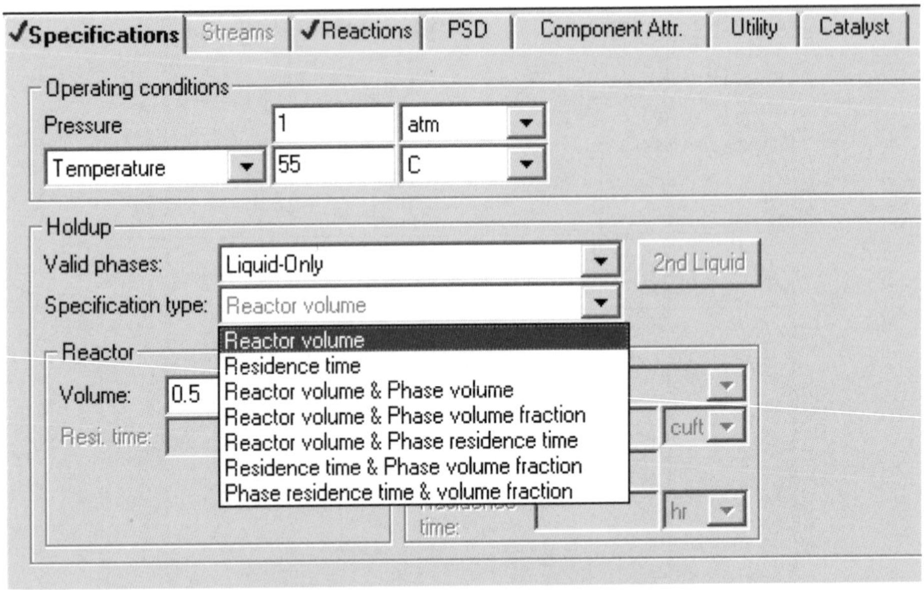

Figure 10.16   Primary RCSTR Input.

rate terms along the reactor length. An overall energy balance is also employed. Details may be found in Levenspiel (1962).

The primary specifications for an RPlug block are shown in Figure 10.17. Each of the reactor types available has an auxiliary set of inputs. For example, in Figure 10.18, a reactor with a constant coolant temperature, the overall heat transfer coefficient

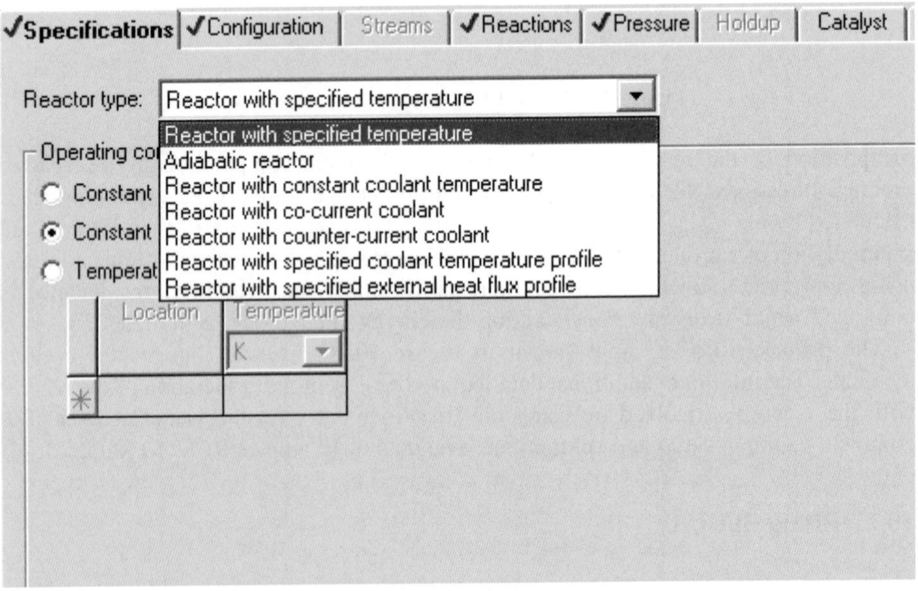

Figure 10.17   RPlug options.

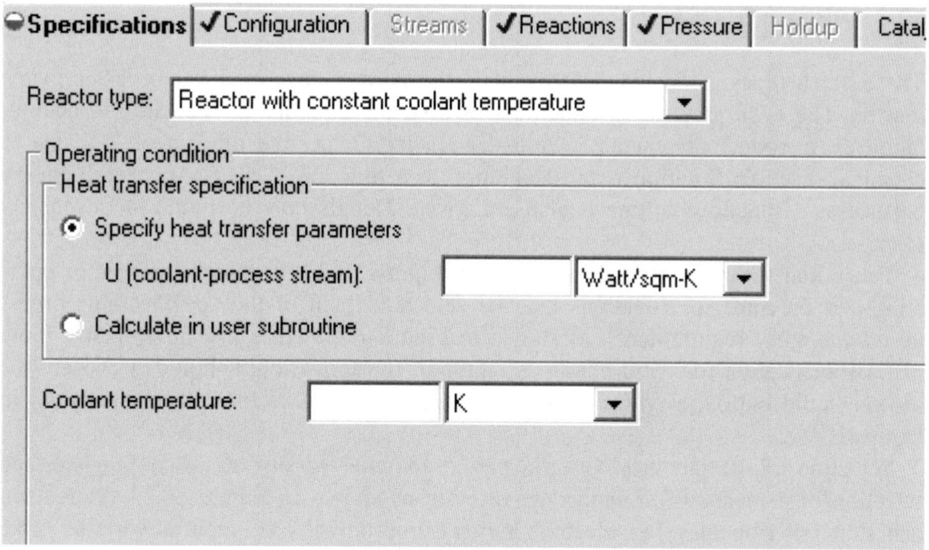

Figure 10.18   Rplug reactor type and heat transfer specifications.

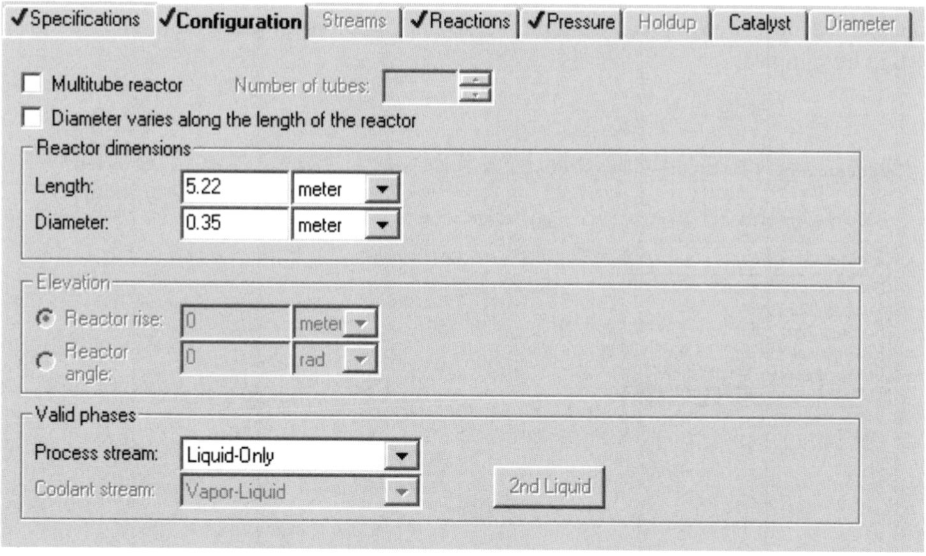

Figure 10.19   RPlug size and layout.

must be provided or, alternatively, calculated in a subroutine. Additionally, the cooler temperature must be provided. When the tab Configuration is selected, Figure 10.19 is presented. Here data dealing with the physical layout of the reactor, as well as the reaction phase are input. Selection of the Pressure tab permits specification of the process stream inlet pressure and various pressure-drop options. The complete example setup and solution appear at Examples/RPlugExample1.

10.8   RBatch BLOCK

The RBatch block is designed to evaluate the performance of a batch or semibatch reactor. The basic describing equations involve the equality of the rate of accumulation of a product or reactant within the reactor to the rate of gain or loss due to chemical reaction. Solution is by integration as a function of time. An overall energy balance as a function of time is also employed. Details may be found in Levenspiel (1962).

The initial setup for Rbatch is shown in Figure 10.20, in which the reactor operating and pressure specifications may be selected. Each of the specifications brings up unique entry requirements, as shown in Figure 10.21. It is also important to note that a specification for valid phases is required. If vapor–liquid–liquid is chosen, the second liquid button provides a means of identifying the second phase, as shown in Figure 10.22.

Selection of the tab Stop Criteria permits a wide variety of possibilities for the selection of a means of stopping the reaction as shown in Figure 10.23. More than one stop criterion may be selected. Various drop-down lists, such as variable type, are available as needed. Figure 10.24 shows the options dealing with batch cycles and control of the number and interval of the results of the simulation. For this example, Figure 10.25 shows mole fractions of each component as a function of time. Additional results, as a function of time, may be found under the tabs Overall and Feed. The complete example setup and solution are available at Examples/ RBatchExample1.

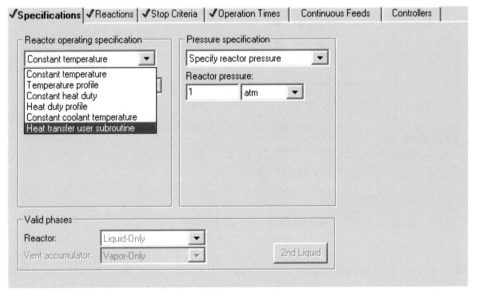

Figure 10.20   RBatch initial setup.

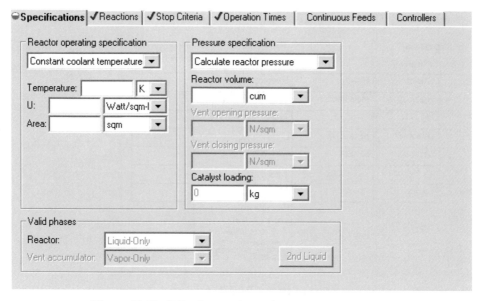

Figure 10.21 RBatch operating and pressure specifications.

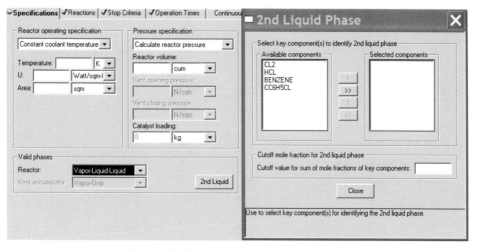

Figure 10.22 RBatch phase specifications.

10.9 WORKSHOPS

Workshop 10.1 Two reactors are connected in series. In the first reactor two reactions take place in series:

$$C_4H_{10} \rightarrow C_2H_6 + C_2H_4$$
$$C_2H_4 + C_6H_6 \rightarrow C_8H_{10}$$

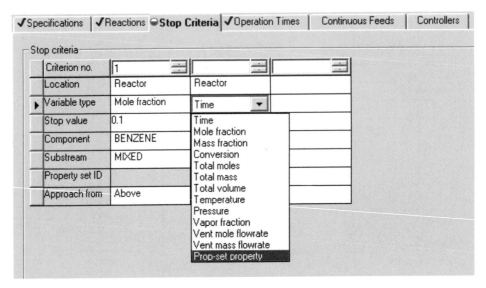

Figure 10.23   RBatch stop criteria.

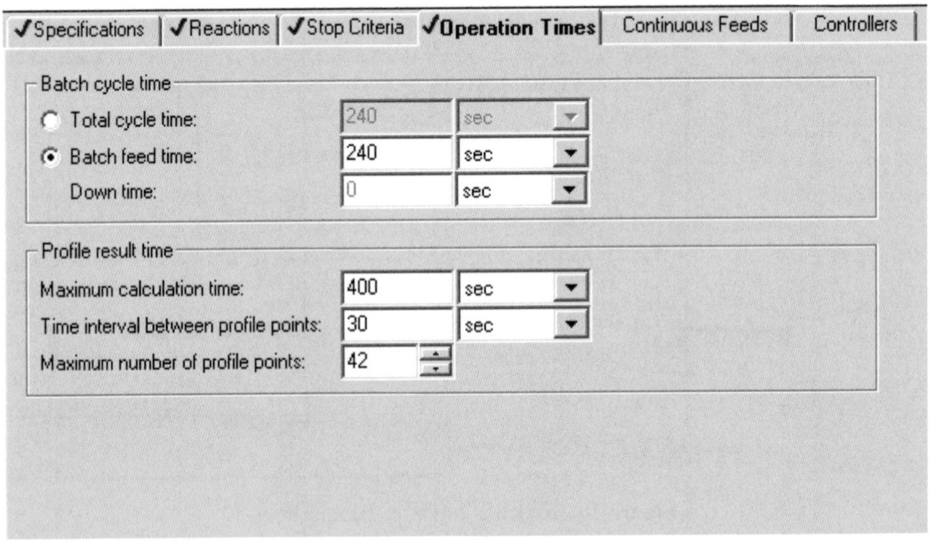

Figure 10.24   RBatch operation time.

The first reactor operates at 300°F and 300 psi. The first reaction conversion, based on nC_4H_{10}, is 50%. The second reaction conversion, based on C_6H_6, is 90%. In the second reactor, which operates at 400°F and 300 psi, the remaining ethylene is completely converted to 1-butene by the reaction

$$2C_2H_4 \rightarrow C_4H_8$$

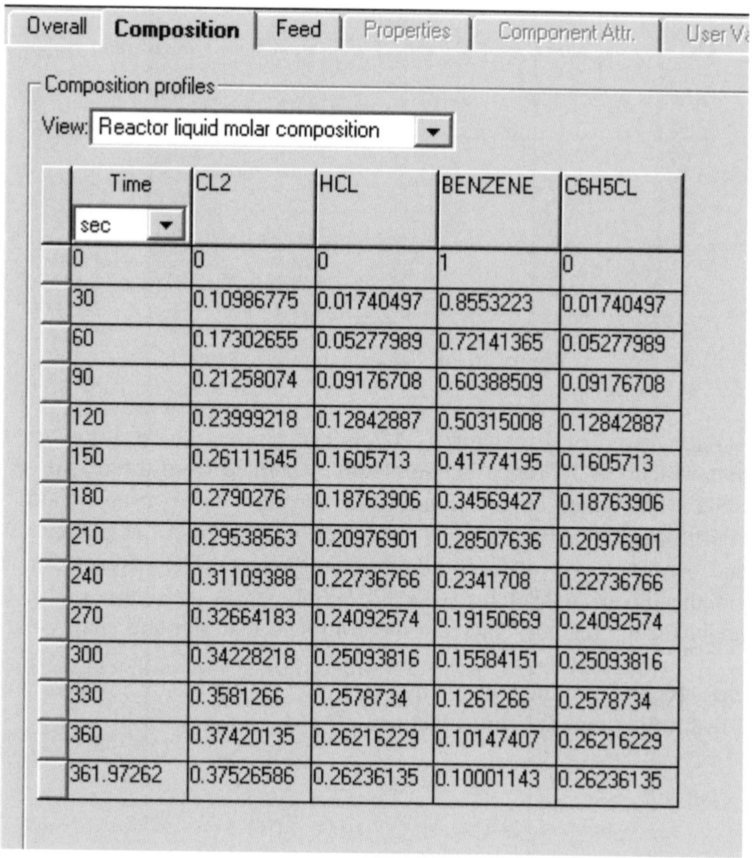

| Overall | **Composition** | Feed | Properties | Component Attr. | User V: |

Composition profiles

View: Reactor liquid molar composition ▼

Time sec ▼	CL2	HCL	BENZENE	C6H5CL
0	0	0	1	0
30	0.10986775	0.01740497	0.8553223	0.01740497
60	0.17302655	0.05277989	0.72141365	0.05277989
90	0.21258074	0.09176708	0.60388509	0.09176708
120	0.23999218	0.12842887	0.50315008	0.12842887
150	0.26111545	0.1605713	0.41774195	0.1605713
180	0.2790276	0.18763906	0.34569427	0.18763906
210	0.29538563	0.20976901	0.28507636	0.20976901
240	0.31109388	0.22736766	0.2341708	0.22736766
270	0.32664183	0.24092574	0.19150669	0.24092574
300	0.34228218	0.25093816	0.15584151	0.25093816
330	0.3581266	0.2578734	0.1261266	0.2578734
360	0.37420135	0.26216229	0.10147407	0.26216229
361.97262	0.37526586	0.26236135	0.10001143	0.26236135

Figure 10.25 Product distribution at various times.

TABLE 10.1 Reactor Train Feeds

	Ethane, C_2H_6	Ethylene, C_2H_4	n-Butane, C_4H_{10}	Benzene, C_6H_6	1-Butene, C_4H_8	Ethylbenzene, C_8H_{10}
Feed 1		10	90			
Feed 2				50		

The feeds for the reactor train, in lbmol/hr, and the component list are given in Table 10.1. The pressure of both feeds is 300 psia, and both are saturated liquids. Calculate the complete material and energy balance.

Workshop 10.2 The following describes a process for making toluene from n-heptane. A fresh feed of 20 lbmol/hr of pure n-heptane at 77°F and 1 atmosphere is combined with a solvent recycle from an extractor and heated to 425°F at 1 atmosphere. The hot stream is fed into a reactor in which the following reaction occurs:

$$C_7H_{16} \rightarrow C_7H_8 + 4H_2$$

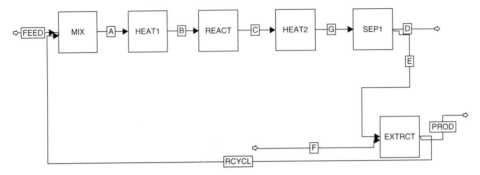

Figure 10.26 Toluene process sketch.

The conversion based on *n*-heptane is 15%. The products of reaction are cooled to 180°F, after which the hydrogen is completely separated from the reactor products in the first separator. A feed of 100 lbmol/hr of benzene at 180°F and 1 atmosphere is combined with the remaining products of reaction to extract the toluene. All of the toluene and benzene leave as the product of the process. The unreacted *n*-heptane is recycled to the mixer. A sketch of the process is given in Figure 10.26. Using the RStoic model for the reactor, calculate the complete material and energy balance.

Workshop 10.3a Production of hydrogen from methanol involves two steps, methanol re-forming and the shift reaction:

$$CH_4O \rightarrow 2H_2 + CO$$

$$CO + H_2O \rightarrow H_2 + CO_2$$

Using an REquil block, find the temperature that maximizes hydrogen formation by re-forming at 1 atmosphere.

Workshop 10.3b Repeat Workshop 10.3a using a RGibbs block. Compare the results.

Workshop 10.3c Using 450°F as the operating temperature, determine the effect of pressure on the reaction. Select the optimal operating conditions. Is there a problem with this analysis?

Workshop 10.3d Choi and Stenger (2002) have suggested that there is a possibility that dimethyl ether and methyl formate may be side products of the re-forming and shift reactions in Workshop 10.3a. Assess this possibility using the RGibbs block at 450°F and a variety of pressures.

Workshop 10.4 The formation of chlorobenzene as described at Examples/RPlug1 Example in Section 10.8 is not limited to chlorobenzene formation but has the possibility of forming both *p*-dichlorobenzene and trichlorobenzene. Belfiore (2003) reports that the relative rates of reaction of the dichloro- and trichlorobenzen relative to the monobenzen are 1/8 and 1/243, respectively. Determine the quantities of the minor products formed for the conditions of RBatchExample1.

Workshop Notes

Workshop 10.1 The choice of serial or parallel reaction is made via a check box in the lower left-hand corner of the reaction setup page.

Workshop 10.2 Initialization is with Aspen Plus defaults (i.e., zeros); however, after one iteration the error messages disappear and the process converges.

Workshop 10.3a Use a sensitivity study, varying the temperature.

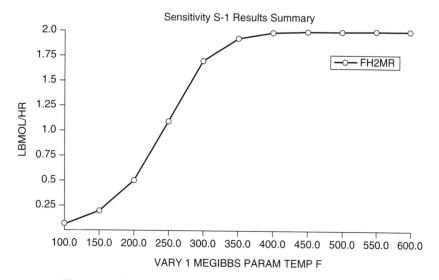

Figure 10.27 Temperature for maximum hydrogen production.

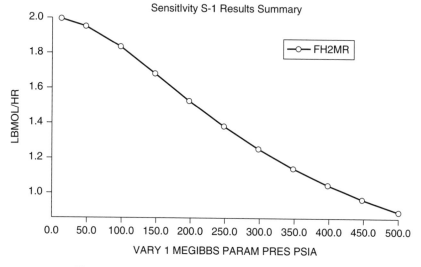

Figure 10.28 Pressure effect on hydrogen production.

Workshop 10.3b   The results are virtually identical to those of Workshop 10.3a. One may observe from Figure 10.27, a plot of the results from a sensitivity study, that 450°F is the optimal operating temperature.

Workshop 10.3c   The effect of pressure shown in Figure 10.28, a plot of the results from a sensitivity study, is pronounced. Based on these results the optimal operating conditions would be 450°F and 14.7 psia. This analysis did not consider pressure effects at other temperatures, and one might find another optimal solution. This point is revisited in Chapter Thirteen.

Workshop 10.3d   Results of the case study show that low pressure favors hydrogen formation and high pressure favors the formation of small quantities of dimethyl ether. Methyl formate formation is negligible.

REFERENCES

Aspen Plus version 7.0, Help, Rate Controlled Reactions.

Belfiore, L. A., *Transport Phenomena for Chemical Reactor Design*, Wiley-Interscience, Hoboken, NJ, 2003, p. 655.

Choi, Y., and Stenger, H., *Fuel Chem. Div. Prep*. 47(2), 724, (2002).

Levenspiel, O., *Chemical Reaction Engineering*, Wiley, Hoboken, NJ, 1962, pp. 100–102, 102–104, and 107–110.

Perry, R. H., and Green, D. W., *Perry's Chemical Engineers' Handbook on CD-ROM*, 7th ed., McGraw-Hill, New York, 1999, pp. 7–11 to 7–13.

CHAPTER ELEVEN

MULTISTAGE EQUILIBRIUM SEPARATORS

Distillation and extraction are the most commonly used separation methods in chemical processing, and models of these are at the heart of Aspen Plus. All the Aspen Plus distillation and extraction models may be found in the model library under the tab Columns.

For a distillation column, simulation implies the existence of a fully configured column (i.e., a number of theoretical stages) and the feed and sidestream locations. When the state of the feeds, and the overhead, bottoms, and sidestream specifications are provided, the model calculates the state of the products, and the reboiler and condenser heat loads, if they were not explicitly specified. For an extraction column the same ideas apply except that no overhead and bottoms specifications are required. There is limited capability in Aspen Plus to design a column directly to yield products of specified composition. Thus, the technique is to search the solution space for a trial column with the characteristics desired, and therefore many rating solutions may be required until one that matches the desired design specifications is obtained.

11.1 BASIC EQUATIONS

All models of distillation and extraction processes involve the solution of the material, equilibrium, and energy balance equations or an applicable subset. All models are based on the analysis of a single stage, such as that shown in Figure 11.1. There are several ways to write the describing equations, based on the choice of independent variables. For the set of equations below, the choices v_i^j and l_i^j, the component i vapor and liquid leaving stage j; f_i^j, the componential feed to stage j; s_i^j, a componential sidestream leaving stage j, T^j and P^j, the temperature and pressure of stage j, and Q^j, the total energy entering or leaving stage j, are used. The phase equilibrium constant

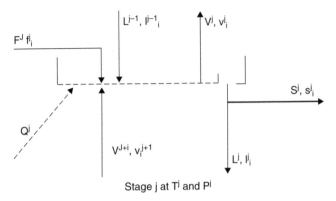

Figure 11.1 Theoretical stage.

K_i^j depends on the stage temperature, pressure, and liquid and vapor mole fractions, x_i^j and y_i^j. Mole fractions can be expressed in terms of componential molar flows; for example, a liquid mole fraction can be calculated by

$$x_i^j = \frac{l_i^j}{\sum_{k=1,m} l_k^j} \qquad (11.1)$$

where m is the number of components. V^j and L^j are auxiliary variables which refer to the total molar vapor and liquid leaving stage j and are calculated by summing the componential flows in the vapor and liquid, leaving a stage as shown in

$$V_j = \sum_{k=1,m} v_k^j$$
$$L_j = \sum_{k=1,m} l_k^j \qquad (11.2)$$

H^j and h^j are the molar enthalpies of the vapor and liquid leaving stage j, which depend on composition, temperature, and pressure.

The material balance for component i, on stage j, is given by

$$f_i^j + v_i^{j-1} + l_i^{j+1} - v_i^j - l_i^j - s_i^j = 0 \qquad (11.3)$$

where s_i^j refers to the molar flow of component i in a sidestream, if it exists. There are nm material balances, where n is the number of theoretical stages and m is the number of components. The introduction of a sidestream adds one degree of freedom since the state of a sidestream is equal to the state of its source (i.e., the phase of the stage from which it is removed), but its flow rate is unknown; however, the composition, temperature, and pressure of the sidestream are known. A specification for a liquid sidestream, such as the ratio of the sidestream to the liquid, removes the extra degree of freedom and permits the added calculation of the componential flows of the sidestream.

The basic equilibrium equations for component i, on stage j, are

$$y_i^j = K_i^j x_i^j \tag{11.4}$$

When the independent variables and equations (11.2) are substituted into (11.4), we obtain

$$v_i^j = K_i^j(v_i^j, l_i^j, T^j, P^j) \frac{l_i^j \sum_{k=1,m} v_k^j}{\sum_{k=1,m} l_k^j} \tag{11.5}$$

There are nm equilibrium equations.

The enthalpy per mole of liquid and vapor flows H^j and h^j refer to the vapor and liquid leaving stage j, respectively. For each stage, one energy balance, given by

$$H_F^j \sum_{k=1,m} f_k^j + H^{j-1} \sum_{k=1,m} v_m^{j-1} + h^{j+1} \sum_{k=1,m} l_k^{j+1} - H^j \sum_{k=1,m} v_k^j - h^j \sum_{k=1,m} l_m^j$$

$$- h^j \sum_{k=1,m} S^j + Q^j = 0 \tag{11.6}$$

may be written, where S^j represents the total molar flow of a sidestream. The enthalpy/mole of a sidestream depends on its source and may be written as either H^j or h^j, depending on the state of the sidestream. If there are no feeds, sidestreams, or heat exchangers associated with stage j, the F, S, and Q terms disappear. There are n energy balance equations.

Models usually do not include stage pressure calculations, and pressures are normally assigned a priori. The summation of equations for a distillation column is as follows:

Material balances	nm
Equilibrium equations	nm
Energy balances	m
Total	$m(2n + 1)$

In the absence of sidestreams or interstage heat exchangers, a count of unknowns for a column with a condenser and a reboiler is as follows:

Componental vapor flows	nm
Componental liquid flows	nm
Stage temperatures	m
Condenser and reboiler duties	2
Total	$m(2n + 1) + 2$

Thus, there are typically two degrees of freedom, which correspond to the column's end specifications. These may be reboiler and condenser duty specifications. Other specifications that can be used, such as product rate and reflux ratio, require minor modifications to the basic equations.

Each sidestream adds one degree of freedom; therefore, it is necessary to provide two end specifications for each sidestream, plus a specification such as a sidestream flow rate.

For extraction columns the equations are identical to those of distillation columns if one recognizes that the vapor terminology now refers to the lighter of the two phases. Additionally, since there are no reboilers or condensers, the degrees of freedom are zero except that one additional degree of freedom is introduced for each sidestream, as in distillation columns.

The set of equations described above lend themselves to solution simultaneously by means of the Newton–Raphson method, such as the algorithm of Naphtali and Sandholm (1971). This method has the advantage of permitting the use of all possible end and sidestream specifications; however, it has the disadvantage of requiring starting estimates for all independent variables.

There are many arrangements of the equations with an accompanying convenient selection of variables that lead to a variety of algorithms to generate a solution. For example, one could employ total flows and mole fractions. There are a group of algorithms that alternately solve the locally linear material balances and then the nonlinear energy balances, adjusting the K values between iterations. The method used in Aspen Plus's Distil block is the inside-out algorithm of Boston and Britt (1974). Details of many of the methods may be found in Seader and Henley (1998).

11.2 THE DESIGN PROBLEM

To simulate a distillation column within a process it is necessary initially to solve a design problem: that is for a given service, to establish the number of stages required and the location of the feeds and sidestreams, and to define suitable end specifications. Since the rigorous distillation models in Aspen Plus do not calculate directly such design parameters as the number of stages and the feed location, it is necessary to explore many possible rating solutions. Prior to the use of a rigorous model, it is useful to estimate the design by less rigorous methods, including the manual McCabe–Thiele method (1925). Prior to any design work, it is imperative that the vapor–liquid equilibrium diagrams and/or liquid–liquid equilibrium diagrams be reviewed. In many cases the data stored in the Aspen Plus database will be suitable, but this may not be the case and a literature search or experimental work may be required.

A brief review of the McCabe–Thiele method is given below. Unlike the models in Aspen Plus, the McCabe–Thiele method can be used to solve the design problem: that is, given the feed composition, flow rate, thermal state, and product composition targets, determine feed location and the number of stages above and below the feed stage. The following features are at the heart of the method.

1. All stages are at steady state and the stage products are at equilibrium.
2. The column operates under equal molal overflow conditions; that is, within a section of a distillation column where there is no feed, no product takeoff, and no interstage heat exchangers, the flow rates of all liquids leaving a stage are identical, as are the flow rates of all vapors.
3. After a solution has been obtained, the reboiler and condenser duties are calculated by a manual energy balance around each device.

The McCabe–Thiele method can also be used to rate columns by a trial-and-error procedure.

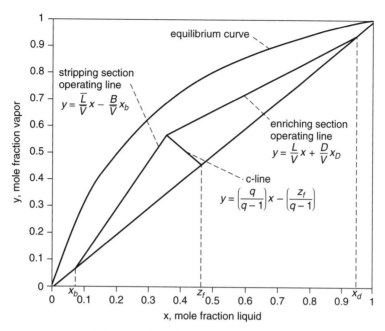

Figure 11.2 McCabe–Thiele diagram.

An overview of the method is given with reference to Figure 11.2, the McCabe–Thiele diagram. The coordinates are the vapor and liquid mole fractions. An equilibrium curve based on experimental data is plotted in the figure. There are operating lines above and below the feed which are derived from the material balance equations in terms of the vapor and liquid mole fractions. The equations of the two operating lines are given in the figure. The two operating lines intersect with the $y = x$ line at the column's respective end compositions and intersect with a feed line that emanates from the feed composition, z_f, located on the $y = x$ line. The slope of the feed line, q, is calculated by

$$q = \frac{H_f - H_F}{H_f - h_f} \tag{11.7}$$

where H_F represents the molar enthalpy of the feed, H_f is the molar enthalpy of saturated vapor, and h_f is the molar enthalpy of saturated liquid. For example, if H_F is a saturated liquid, the value of q is zero, which represents a horizontal line on the McCabe–Thiele diagram.

The number of theoretical stages is obtained by drawing a vertical line from an operating line to the equilibrium curve. This is equivalent to finding the equilibrium compositions for a stage. A horizontal line from the equilibrium curve back to the operating line gives the compositions of the passing streams (i.e., the material balance). This is repeated from one end of the column to the other.

An important feature is the ability to determine the minimum reflux ratio, L/V, which occurs at an infinite number of stages. This is known as the *pinch point* and usually occurs when a line drawn from the overhead composition specification is connected

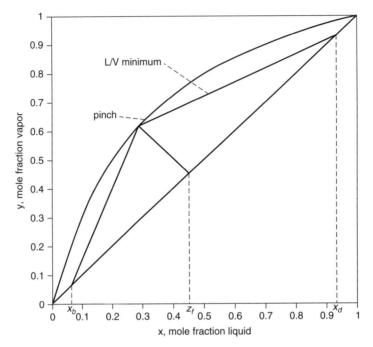

Figure 11.3   Minimum reflux.

to the feed line/q-line intersection, as shown in Figure 11.3. If the equilibrium curve has an unusual shape, as in a methanol–water system, a different pinch point may be located, as shown in Figure 11.4. A rule of thumb commonly used for the design of distillation columns is that the operating reflux should be between 1.25 and 2.0 times the minimum. An estimate of the minimum number of stages required, which occurs at total reflux, is shown in Figure 11.5.

If a sidestream is required, a third operating line is introduced, as shown in Figure 11.6. This operating line starts at the feed/q-line intersection and has a slope L'/V'. After determining the operating reflux ratio, a McCabe–Thiele diagram can be constructed and theoretical stages stepped off as described above. A detailed McCabe–Thiele example follows.

11.3 A THREE-PRODUCT DISTILLATION EXAMPLE

A 20 mol % ethanol–water solution is to be distilled. Two products, 80 mol % and 40 mol %, are required. The feed is available as a saturated liquid. The composition of the bottom product should not exceed 0.25 mol % ethanol. The external reflux ratio is to be four times the minimum. The shape of the equilibrium curve necessitates a relatively high reflux ratio, The sidestream draw-off is 10 lbmol/hr. Using a basis of 100 lbmol/hr, determine the number of theoretical stages required, the feed location, the sidestream location, and the quantities of the products. The experimental data of Kirschbaum and Gerstner (1939) are to be used.

The data are plotted in Figure 11.7, as is the minimum L/V of 0.5, which corresponds to a minimum external reflux ratio L/D of 1.0. The enriching section operating line's

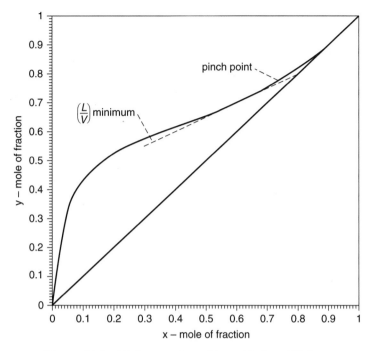

Figure 11.4 Minimum reflux with nonideal equilibrium.

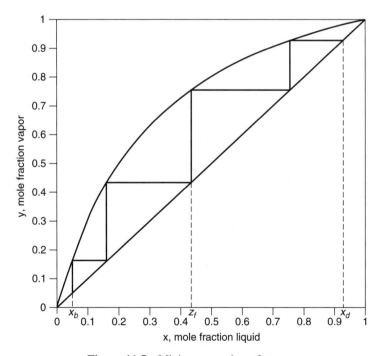

Figure 11.5 Minimum number of stages.

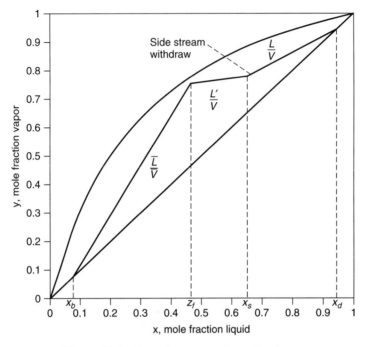

Figure 11.6   Operating lines with a sidestream.

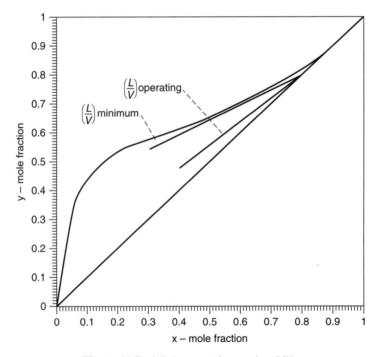

Figure 11.7   Minimum and operating L/V.

L/V of 0.8 is derived from the external reflux ratio specification, which is also shown and corresponds to an operating external reflux ratio of 4.

An overall material balance yields the following results:

Distillate, D = 24.8 lbmol/hr, $x_D = 0.8$
Sidestream, S = 10 lbmol/hr, $x_S = 0.4$
Bottoms, B = 65.2 lbmol/hr, $x_B = 0.25$

The various liquid and vapor flow rates throughout the column, based on equal molar overflow, are given in Figure 11.8, from which the operating line slopes are derived:

Enriching section: $L/V = 99.2/124 = 0.8$
Sidestream section: $L'/V' = 89.2/124 = 0.72$
Stripping section: $L''/V'' = 189.2/124 = 1.53$

The preliminary design is given in Figure 11.9. There are a total of eight theoretical equilibrium contacts, one of which is the reboiler, as well as a total condenser. The feed is on stage 7 and the sidestream is on stage 6, counting from the top down. This

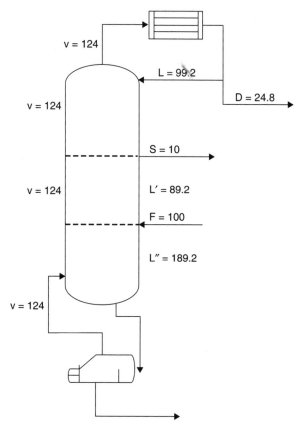

Figure 11.8 Three-product distillation example with equal molal overflow.

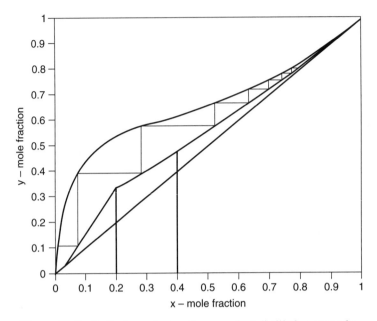

Figure 11.9 Preliminary design: three-product distillation example.

is one of several possible solutions, since there is a trade-off between the number of stages, the reflux ratio, the equipment cost, and energy considerations.

11.4 PRELIMINARY DESIGN AND RATING MODELS

11.4.1 DSTWU

The DSTWU model is applicable to both binary and multicomponent systems and was developed for systems that exhibit approximately constant relative volatility. The DSTWU model is based on the Gilliland correlation shown in Figure 11.10. This correlation was developed from the operating data of many columns and relates the minimum and operating reflux ratios, R_{min} and R, to the minimum and actual number of theoretical stages, N_{min} and N. Gilliland's data were correlated by Molokanov et al. (1972), which resulted in an analytical version of the Gilliland correlation:

$$\frac{N - N_{min}}{N + 1} = 1 - \exp\left(\frac{1 + 54.4X}{11 + 117.2X} \frac{X - 1}{X^{0.5}}\right) \tag{11.8}$$

where

$$X = \frac{R - R_{min}}{R + 1}$$

The minimum reflux ratio is obtained by applying Underwood's (1946) method, which is applicable to binary or multicomponent systems. In the latter case it is necessary to select light and heavy key components and to solve for the roots of

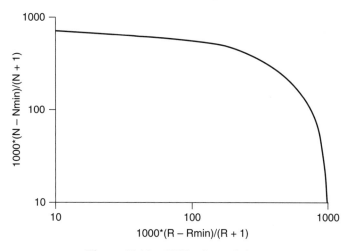

Figure 11.10 Gilliland correlation.

$$\sum_i \frac{\alpha_i z_i}{\alpha_i - \theta} = 1 - q \tag{11.9}$$

Here α_i is the relative volatility of component i, z_i is the mole fraction in the feed of component i, q is the thermal state of the feed, using the McCabe–Thiele definition, and θ is a root of the equation. To calculate the minimum reflux ratio, the value of θ, which lies between the light and heavy keys, is substituted into the equation

$$R_{\min} + 1 = \sum_i \frac{\alpha_i x_i^D}{\alpha_i - \theta} \tag{11.10}$$

where x_i^D refers to the mole fraction of component i in the distillate.

The minimum number of stages is calculated using Winn's (1958) equation, an improvement on Fenske's (1932) equation, where the usual definition of relative volatility is replaced by

$$K_i = \beta_i (K_R)^{\theta_i} \tag{11.11}$$

and the parameters β_i and θ_i are determined from the equilibrium data for each component. Winn's equation is

$$N_{\min} = \frac{\ln \left[(x_D/x_B)_{LK} \, (x_B/x_D)_{HK}^{\theta_{LK}} \right]}{\ln \beta_{LK}} \tag{11.12}$$

The procedure for estimating the distribution of nonkey components is as follows. Assuming that the equation

$$\frac{d_i}{b_i} = \frac{d_r}{b_r} \alpha_{i,r} \tag{11.13}$$

applies where d_i, b_i, d_r, and b_r are the distillate and bottoms componential flows of components i and r, and $\alpha_{i,r}$ is their relative volatility, and when equation (11.13) is combined with the overall componential material balance equation,

$$f_i = d_i + b_i \qquad (11.14)$$

where f is the feed of component i, the equations

$$b_i = \frac{f_i}{1 + (d_r/b_r)\,\alpha_{i,r}} \qquad (11.15a)$$

$$d_i = \frac{f_i\,(d_r/b_r)\,\alpha_{i,r}}{1 + (d_r/b_r)\,\alpha_{i,r}} \qquad (11.15b)$$

are obtained.

An example of the use of DSTWU is provided at Chapter Eleven Examples/DSTWU Example. The model's primary input is shown in Figure 11.11. Selection of the Calculation Options tab enables a user to specify a table of reflux versus the number of theoretical stages.

11.4.2 Distl

The Distl model is applicable to both binary and multicomponent systems and makes use of the equal molal overflow and constant relative volatility approximations. The model employs Edmister's method and rates single-feed, two-product columns. Details may be found in Seader and Henley (1998).

The primary input form is shown in Figure 11.12. The DSTWU example was used as input to the Distl model. Results are given at Examples/DistlExample.

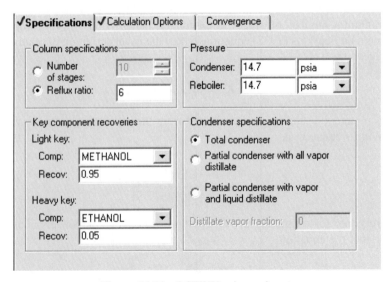

Figure 11.11   DSTWU primary input.

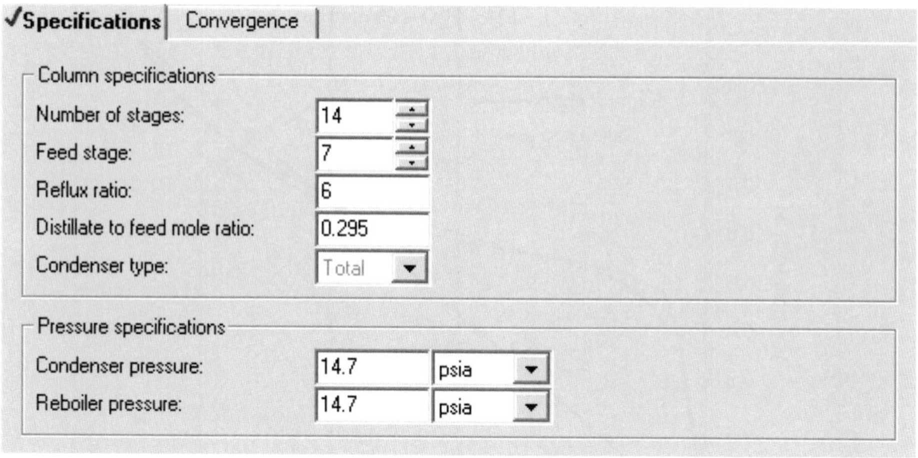

Figure 11.12   Distl primary input.

11.5   RIGOROUS MODELS

After developing a preliminary design, it should be possible to develop a rating model of the system of interest. If the system is nearly ideal, or the phases can be represented by an equation of state, the Aspen Plus database is probably suitable. If, on the other hand, the system is sufficiently nonideal that an activity coefficient equation is required, it is advisable to assess the equilibrium data within Aspen Plus by comparing them with experimental data. If the data do not compare favorably, fit suitable data to an activity coefficient model. If the Aspen Plus database is lacking, acquire suitable equilibrium data either by searching the literature. developing data experimentally, or estimating data using Unifac. Once suitable data have been found, they should be fit to an activity coefficient model.

As an example, the vapor–liquid equilibrium data of Wilson and Simons' (1952) for the isopropanol–water system at 95 mmHg is to be considered for a distillation application. The experimental data were compared against results calculated using a property analysis run "points along a flash curve" using Aspen Plus's stored data fit with the Wilson equation. Figure 11.13 shows a plot of both data sets.

The usefulness of Aspen Plus's stored data should be assessed by an engineer in light of the precision required for the application. For example, If the application required that the product composition not be in the region approaching the azeotrope, the Aspen Plus data would suffice since the data calculated differ very little from the experimental data. If however, the proposed column required that the product be near the composition of the azeotrope, Figure 11.13 shows that the azeotrope composition calculated is about 0.05 mole fraction higher than the experimental value. This would produce serious problems, for example, if this column were part of an azeotropic distillation scheme involving two columns operating at different pressures. The quality of the 95 mmHg data calculated from the stored Wilson parameters is suitable since the data that were fit ranged between 77 and 212°F, as can be seen from Figure 11.14, and the experimental data range from a low of about 35°C (95°F) to a high of 49°C (120°C), but unfortunately, do not accurately model the composition of the azeotrope.

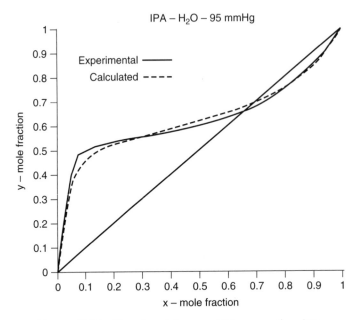

Figure 11.13 Experimental versus Wilson equation data.

If no experimental data are available for a system, or data cannot be developed in a timely manner, the Unifac group contribution method may be used to estimate the data. Although these data may be used for preliminary design, it would be wise to set in motion means for experimental verification. For example, Figure 11.15 shows a comparison of experimental and Unifac estimated vapor–liquid equilibrium for an isopropyl alcohol–water system at 95 mmHg. The prediction of the azeotrope composition is remarkable, but the lower composition range is acceptable at best, but again, use of these data is application driven.

11.5.1 RadFrac

RadFrac is the rigorous distillation modeling workhorse of Aspen Plus. It is based on the inside-out formulation of Boston and Sullivan (1974), which employs a reorganization of the basic equations and the sequence of calculation. A detailed description of the method may be found in Seader and Henley (1998). For problems in which convergence difficulties are experienced, five variations of the basic algorithm are available.

The first example of RadFrac use here is the same as that used for DSTWU and Distl. Since RadFrac is a rating or simulation program, the necessary specifications are obtained from the results of the DSTWU example (i.e., feed stage location, reflux ratio, and distillate rate). The primary input form for RadFrac is shown in Figure 11.16. Most of the remaining process specifications are straightforward and accessed from the tabs available. The tabs are Streams, in which the feeds and products are specified; Pressure, in which the stage pressures are specified; Condenser, in which the condenser details are specified; and Reboiler, in which the reboiler details are specified. If in Figure 11.16 the reboiler specified is thermosyphon, the Thermosyphon Config and

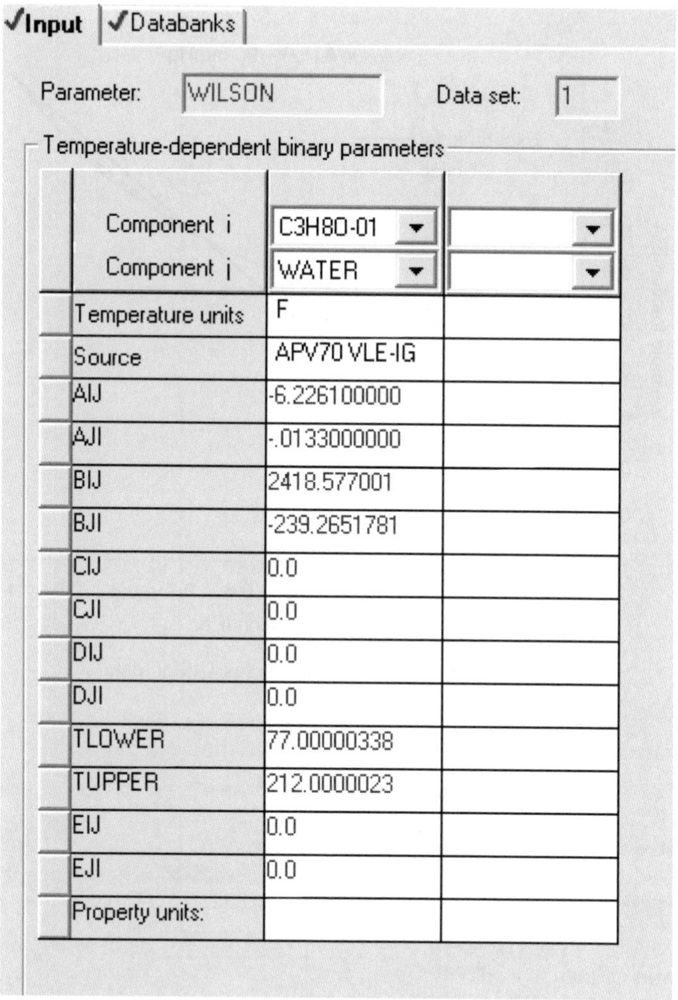

√Input	**√Databanks**			

Parameter: WILSON   Data set: 1

Temperature-dependent binary parameters

Component i	C3H8O-01 ▼	▼
Component j	WATER ▼	▼
Temperature units	F	
Source	APV70 VLE-IG	
AIJ	-6.226100000	
AJI	-.0133000000	
BIJ	2418.577001	
BJI	-239.2651781	
CIJ	0.0	
CJI	0.0	
DIJ	0.0	
DJI	0.0	
TLOWER	77.00000338	
TUPPER	212.0000023	
EIJ	0.0	
EJI	0.0	
Property units:		

Figure 11.14   Stored isopropyl alcohol–water Wilson parameters.

Rebolier tabs become available. When the Thermosyphon Config tab is selected, the thermosyphon reboiler baffle configuration may be selected, as shown in Figure 11.17. When the tab Reboiler specification is chosen, Figure 11.18 appears. This permits various reboiler specifications to be made for either kettle or thermosyphon reboilers. If the Reboiler Wizard button is pushed, the Reboiler Wizard display, on the right of Figure 11.18, appears and enables details of an associated heat exchanger to be entered, including a link to the Heatx block. Results for this simulation may be found at Example/RadFracExample1.

The RadFrac block has an associated design specification function. In the example above, the DSTWU solution shows a 0.95 recovery fraction of methanol; however, the rigorous RadFrac solution shows that the recovery is 0.937. One possibility for improving the recovery is to vary a column specification, in this case either the distillate rate or the reflux ratio. Use of the design specification, varying the distillate rate,

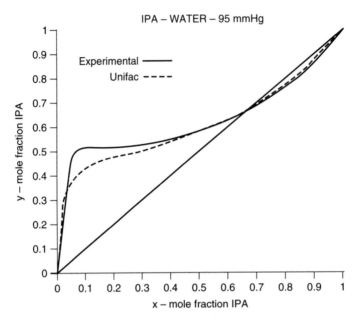

Figure 11.15 Experimental versus Unifac data.

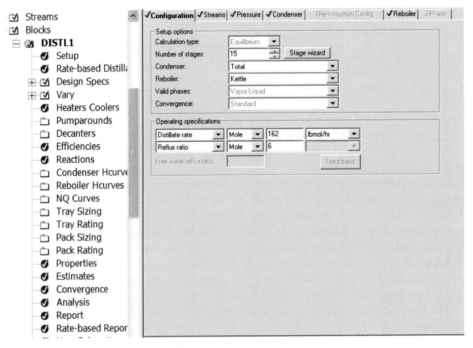

Figure 11.16 Primary RadFrac input.

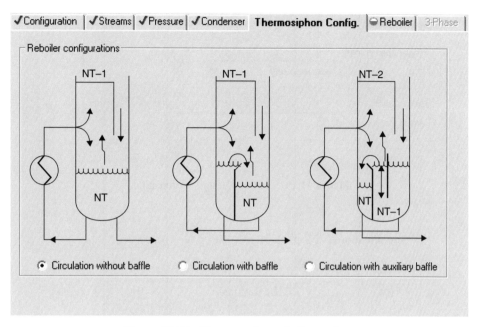

Figure 11.17   Thermosyphon reboiler options.

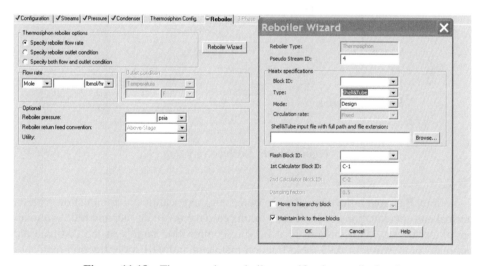

Figure 11.18   Thermosyphon reboiler specifications and wizard.

is illustrated at RadFracExample2. Figure 11.16 shows a list of options under the configured block named Distl1. The entry Design Specs, followed by Object Manager, New, and Mole recovery target of 0.95, will create the specification, which for this example is shown in Figure 11.19. The selections of Components and Feed/Product Streams allow completion of the specification. Figure 11.20, created by using the Vary entry, shows the bounds within which to search for a result. Results are given at Examples/RadFracExample2.

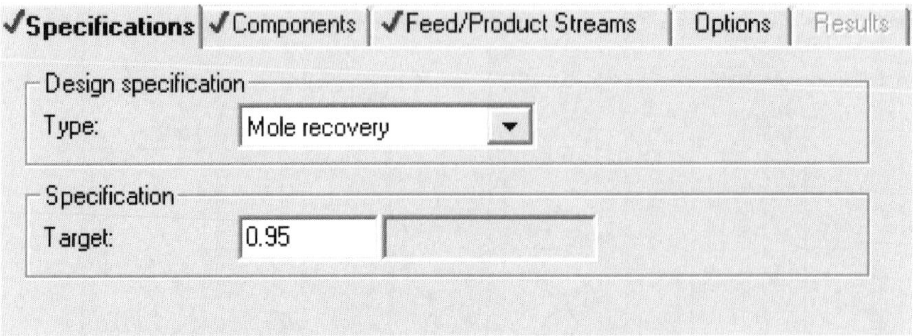

Figure 11.19 RadFrac design specification.

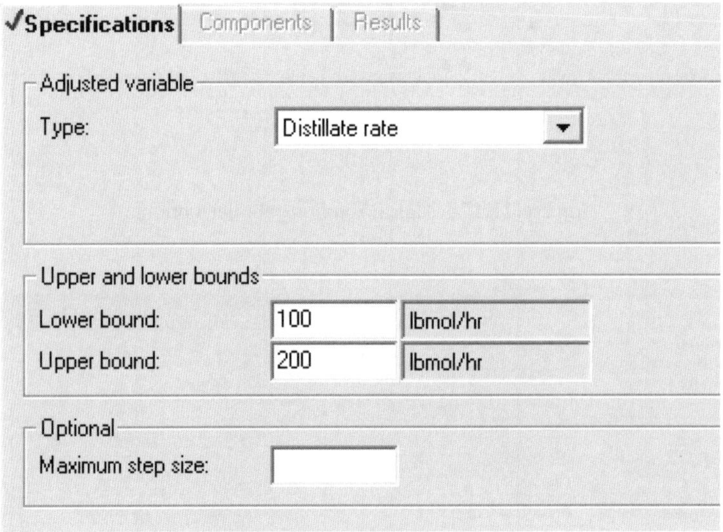

Figure 11.20 RadFrac vary form.

RadFracExample3 illustrates the use of design specifications in conjunction with the normal end specifications of RadFRac. In this case the distillate rate and reflux ratio are permitted to vary, with the aim of achieving composition targets of the distillate and bottoms composition of 0.95 mole fraction methanol and 0.01 mole fraction methanol, respectively. The solution shows that for the limits of the perturbed variables the bottoms composition is obtained and the overhead specification is close and probably acceptable. Note that the Aspen Plus results show many errors, but in a practical sense they are meaningless. These results are obtained by using two implementations of the Design Spec and Vary functions. The final values of the reflux ratio and the distillate rate are given in the report.

The remaining design questions are: Are the number of stages nearly optimal? Where is the best feed location? RadFrac addresses these questions by selecting NQ Curves under the list of options for the entry Distl1. The setup of this capability is shown in Figure 11.21 and is illustrated at Examples//Example3a. In this example, many errors,

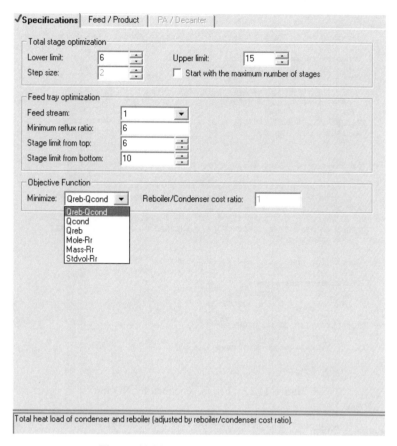

Figure 11.21 NQ curves specifications.

some of which are shown in Figure 11.22, are generated. The various reasons for the errors include limitations at the bounds of perturbed variables. In this case the optimal feed location and number of stages required were determined correctly from DSTWU and are given in the base case solution. This will usually not occur unless the species involved form relatively ideal mixtures.

Aspen Plus's RadFrac block provides many detail design functions, which are accessible through the Tray and Packing Sizing and Rating folders under the RadFrac block. These functions are based on vendor-provided information or well-known methods described in the literature. Details of the calculations, correlations, and so on, are given in Aspen Plus's Help documentation, Appendix A. An example of the tray design input is given in Figure 11.23. Note that five different tray types are available. In this example the sizing function was used for three sections of the column: enriching, feed, and stripping. Results are shown at Examples/Example3b. When using RadFrac for rating or simulation, the design specifications should be removed and operating specifications such as reflux ratio and distillate rate should be used. Additionally, the tray or packing design specifications should be removed and replaced with the equivalent tray and packing rating functions. An example of such an implementation is shown at Examples/Example3c. Figure 11.24 shows an example of the packing input.

```
    Iteration =  2  Feed Stage =  6  Objective Function =  0.37617E+08

    Convergence iterations:
       OL   ML   IL       Err/Tol
        1    4   12        73.239
        2    2    4        3.9229
        3    2    3        0.96662
 ** ERROR
    DESIGN SPEC IS NOT SATISFIED BECAUSE ONE OR MORE MANIPULATED
    VARIABLE IS AT ITS BOUND.


    Iteration =  3  Feed Stage =  4  Objective Function =  0.37618E+08

    Feed tray optimization converged for NSTAGE =   8
      Optimum feed stage =  5  Objective Function =  0.37611E+08
 *  WARNING
    NQ CURVE "1" STOPPED AT STAGE "8" SINCE THERE IS NO SIGNIFICANT
    CHANGE IN OBJECTIVE FUNCTION WHEN INCREASING NUMBER OF STAGES.


   ** Converging the base case for NSTAGE = 15 **

    Convergence iterations:
       OL   ML   IL       Err/Tol
        1    5   14        0.38203E-03

 *** END GENERATION OF NQ CURVES ***

->Simulation calculations completed ...
```

Figure 11.22   Errors generated by NQ run.

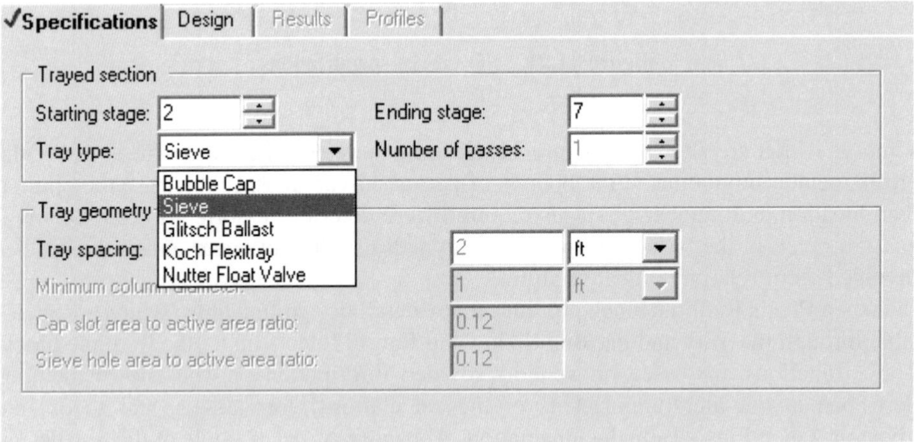

✓**Specifications** | Design | Results | Profiles |

Trayed section
Starting stage: 2   Ending stage: 7
Tray type: Sieve ▼   Number of passes: 1

    Bubble Cap
    Sieve
    Glitsch Ballast
Tray geometry   Koch Flexitray
Tray spacing: Nutter Float Valve   2   ft   ▼
Minimum column diameter: 1   ft   ▼
Cap slot area to active area ratio: 0.12
Sieve hole area to active area ratio: 0.12

Figure 11.23   Tray design input.

11.5.2 Extract

Extract is a rigorous extraction column simulation block. The material, energy, and equilibrium equations given in equations (11.1) through (11.6) apply with the understanding that the vapor-phase nomenclature is fully appropriate for interpretation as the second liquid. Because there are no phase changes, there are rarely large energy

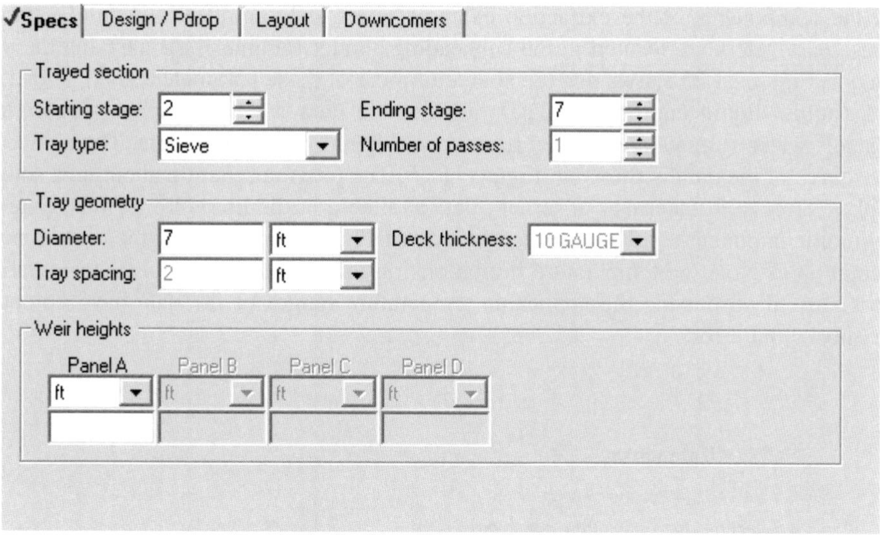

| ✓Specs | Design / Pdrop | Layout | Downcomers |

Trayed section

Starting stage: 2   Ending stage: 7

Tray type: Sieve   Number of passes: 1

Tray geometry

Diameter: 7 ft   Deck thickness: 10 GAUGE

Tray spacing: 2 ft

Weir heights

| Panel A | Panel B | Panel C | Panel D |
| ft | ft | ft | ft |

Figure 11.24   Packing input.

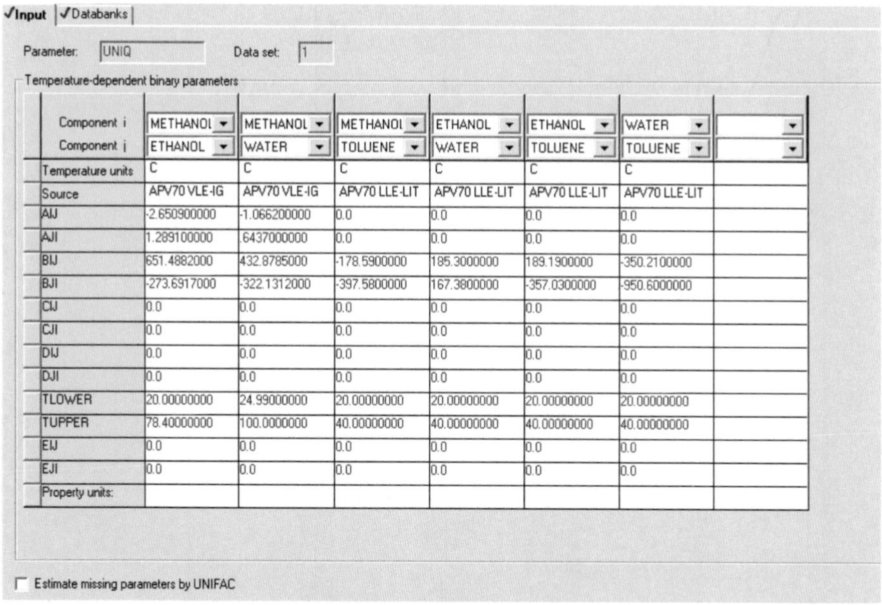

| ✓Input | ✓Databanks |

Parameter: UNIQ   Data set: 1

Temperature-dependent binary parameters

Component i	METHANOL	METHANOL	METHANOL	ETHANOL	ETHANOL	WATER	
Component j	ETHANOL	WATER	TOLUENE	WATER	TOLUENE	TOLUENE	
Temperature units	C	C	C	C	C	C	
Source	APV70 VLE-IG	APV70 VLE-IG	APV70 LLE-LIT	APV70 LLE-LIT	APV70 LLE-LIT	APV70 LLE-LIT	
AIJ	-2.650900000	-1.066200000	0.0	0.0	0.0	0.0	
AJI	1.289100000	.6437000000	0.0	0.0	0.0	0.0	
BIJ	651.4882000	432.8785000	-178.5900000	185.3000000	189.1900000	-350.2100000	
BJI	-273.6917000	-322.1312000	-397.5800000	167.3800000	-357.0300000	-950.6000000	
CIJ	0.0	0.0	0.0	0.0	0.0	0.0	
CJI	0.0	0.0	0.0	0.0	0.0	0.0	
DIJ	0.0	0.0	0.0	0.0	0.0	0.0	
DJI	0.0	0.0	0.0	0.0	0.0	0.0	
TLOWER	20.00000000	24.99000000	20.00000000	20.00000000	20.00000000	20.00000000	
TUPPER	78.40000000	100.0000000	40.00000000	40.00000000	40.00000000	40.00000000	
EIJ	0.0	0.0	0.0	0.0	0.0	0.0	
EJI	0.0	0.0	0.0	0.0	0.0	0.0	
Property units:							

☐ Estimate missing parameters by UNIFAC

Figure 11.25   Uniquac parameters and data sources.

swings in extraction columns, and thus they operate virtually isothermally. There are no condensers or reboilers, and therefore the number of degrees of freedom is equal to the number of available equations, and no operating specifications are required. The Extract block is a rating or simulation model, and its use requires the specification of feed rates, states, and locations as well as the number of equilibrium stages.

When preparing an extraction column simulation, a review of the available data in Aspen Plus's databases is imperative. Figure 11.25 shows the Uniquac parameters for the components of the extraction example given at Examples/ExtractionExample. These parameters are located in the subheadings under Parameters/Binary Interaction/ Uniq-1. Figure 11.25 shows that the source of most of these parameters is APV70LLE-LIT (liquid–liquid equilibrium data) but that the data for the methanol–water and ethanol–water systems are derived from vapor–liquid equilibrium data. The choice of the source of the data is shown in Figure 11.26. The previous caution about data source applies here. Are quaternary or ternary data available in the literature for this system? Do multicomponent results generated by Aspen Plus's stored data agree with experimental data? Note the temperature limitations on the data in Figure 11.25. Simulation work carried on outside the applicable temperature ranges of the data have potential for substantial error.

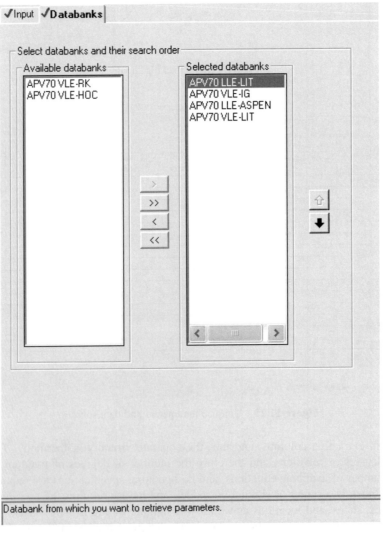

Figure 11.26 Data source choices.

11.6   BatchSep

Batch distillation is commonly used in processing specialty chemicals and pharmaceuticals and has a substantial role in the chemical industries. Aspen Plus's batch distillation block is BatchSep. The equations describing batch distillation are very similar to those describing steady-state distillation; however, the material and energy balances are time varying, and therefore differential equations are required. An example of a differential material balance analogous to equation (11.3) is

$$\frac{d(x_i^j \rho M^j)}{dt} = z_i^j F^j + y_i^{j+1} V^{j+1} + x_i^{j-1} L^{j-1} - y_i^1 V^j - x_i^j L^j - x_i^j S^j \qquad (11.16)$$

Here the alternative form of the material balance is used, with the independent variables being mole fractions and total flows. If M^j is the volumetric stage holdup and ρ is the molar density, equation (11.16) results. Similarly, a differential energy balance analogous to equation (11.6) is

$$\frac{d(\rho M^j c_p T)}{dt} = H_F^j F^j + HV^{j+1} + hL^{j-1} - HV^j - hL^j - hS^j + Q \qquad (11.17)$$

For an n-component column, there are n differential material balances and one differential energy balance per stage. Additionally, n nonlinear algebraic equations describing phase equilibrium and the auxiliary summation of vapor and liquid mole fractions equations are also required per stage. This large set of combined differential and nonlinear equations can be solved numerically. Initial conditions are required for all the variables described in differential equation terms. Targets during the distillation, such as overhead temperature, are used to identify the beginning and end of a cut, and receivers must be defined for the collection of each cut produced.

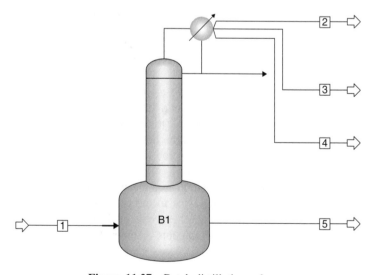

Figure 11.27   Batch distillation column.

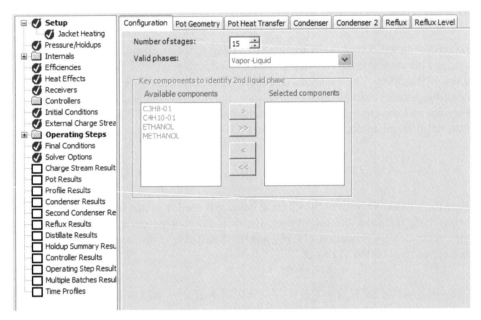

Figure 11.28 BatchSep basic setup.

An example of a batch distillation application is given at Examples/BatchSep Example. Figure 11.27 shows a sketch of a batch distillation column. Stream 1 is the batch charge, streams 2, 3, and 4 are the products (cuts), and stream 5 is the pot contents at the conclusion of the distillation. Each of the product streams has an associated receiver. BatchSep applications require Aspen Plus release 7.1 or higher.

The Setup form for the BatchSep block is shown in Figure 11.28. Note the possibility of two liquid phases in equilibrium with a vapor phase. Each of the available tabs enables definition of the physical dimensions of the still as well as the column's overhead specification, such as the reflux ratio. The Jacket Heating tab enables the choice and value of the bottoms specification. The tab Receivers permits definition of the number of receivers and the association of each product stream with a particular receiver. The Side Draws tab permits the entry of all necessary side draw information.

The Controllers folder facilitates the use of a simulated PID-type controller during batch distillation. Figure 11.29 shows the two tabs that permit a user to choose a variable to be controlled and the controller settings that operate the controller. This capability enables one to simulate the real-time behavior of the batch still.

The input forms for the Initial Conditions tab are illustrated in Figure 11.30. Both the Main and Initial Charge tabs are shown.

Control of the products of the distillation is done with the Operating Steps entry. In this example there are two operating steps to the process. Figure 11.31 shows the setup of Stepone. When all the steps have been executed, the Final Conditions entry permits the choice of disposition of the contents of the reflux drum, the packing, and the pot.

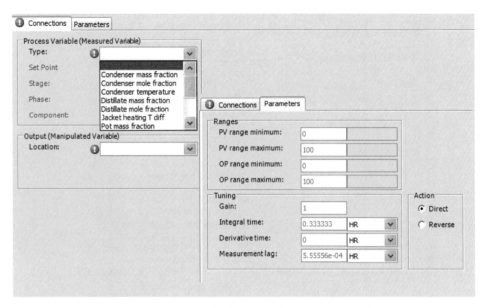

Figure 11.29 Batch distillation column controller setup.

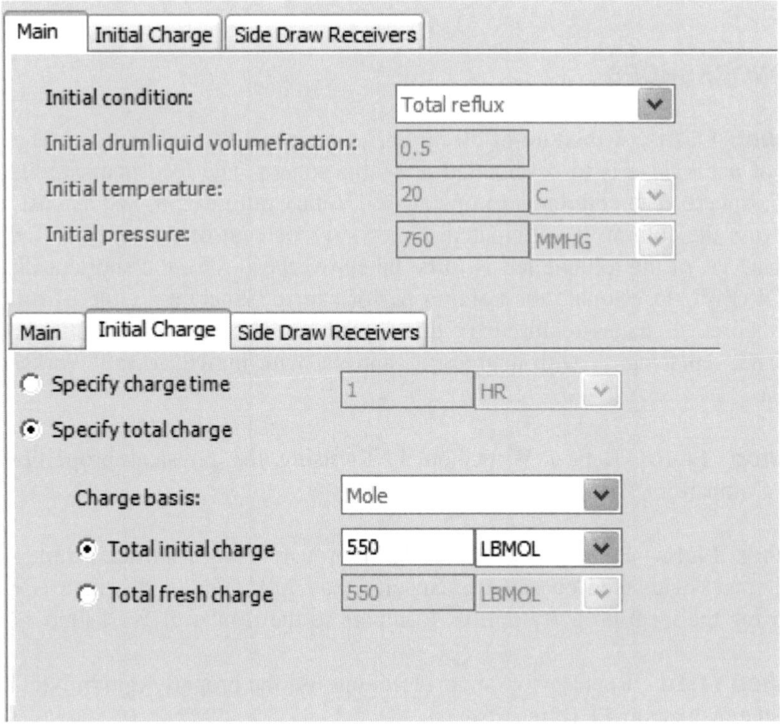

Figure 11.30 Initial conditions for BatchSep.

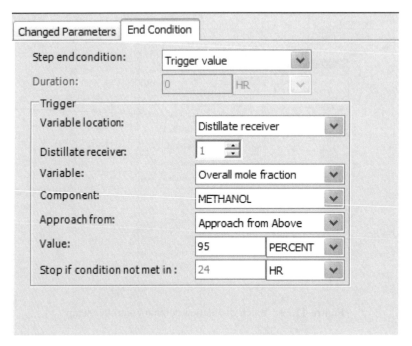

Figure 11.31   Operating step input, Stepone.

11.7 WORKSHOPS

Workshop 11.1a   A mixture of 50 mol/hr of benzene, 50 mol/hr of toluene, and 50 mol/hr of paraxylene is to be distilled at 1 atmosphere. The feed is a saturated liquid at 1 atmosphere. The column is to operate at a reflux ratio of 2.8, and a 1-psi pressure drop across the column is anticipated. Ninety-five percent of the benzene fed and no more than 5% of the toluene fed is to be taken overhead. Use the shortcut distillation block DISTWU to estimate the column performance. Generate a table of reflux ratio versus theoretical stages required for this separation. Create a mixed unit set derived from ENG, replacing °F with degrees °C and psi with mmHg for all Workshop 11.1 variations.

Workshop 11.1b   Repeat Workshop 11.1a using the physical properties option NRTL. Compare to Workshop 11.1a.

Workshop 11.1c   Solve Workshop 11.1a rigorously using Radfrac. Employ 12 theoretical stages with feed on stage 6. Specify that 50 mol/hr of distillate is required. Solve using the ideal property option. Compare to the results of Workshop 11.1a.

Workshop 11.1d   Repeat Workshop 11.1c, but use the property option NRTL. Compare to the Workshop 11.1c results.

Workshop 11.2a   A mixture of 50 mol/hr of methanol, 50 mol/hr of ethanol, and 50 mol/hr of water is to be distilled at 1 atmosphere. The feed is a saturated liquid at

1 atmosphere. The column is to operate at a reflux ratio of 1.2, and a 1-psi pressure drop across the column is anticipated. Ninety-five percent of the ethanol fed, and no more than 5% of the water fed, is to be taken overhead. Use the shortcut distillation block DISTWU with the physical properties option Wilson to estimate the column performance. Generate a table of reflux ratio versus theoretical stages required for this separation. Create a mixed unit set derived from ENG, replacing °F with °C and psi with mmHg for all Workshop 11.2 variations.

Workshop 11.2b   Solve Workshop 11.2a rigorously using the block RadFrac. Specify that 50 mol/hr of bottoms is required. Use 15 theoretical stages with feed on stage 10. Plot the liquid compositions for the column. What do you observe? Compare to the results of Workshop 11.2a.

Workshop 11.2c   Repeat Workshop 11.2b rigorously. Specify that 100 mol/hr of distillate is required. Use 11 theoretical stages with feed on stage 7. Compare to the results of Workshop 11.2b.

Workshop 11.3a   The data for this problem, from Udovenko and Mazanko (1964), are available in the file DataEleven-3bkp in the Workshops folder. Create a set of units, NEWSET, which employs pressure in mmHg and ΔP in atm, copied from the data set MET. Fit the data to the NRTL equation, assigning 0.0 to the $A_{i,j}$ parameters and assigning 0.2 to each C parameter. Produce a triangular plot of the data.

A mixture of 500 lb/hr of benzene and 500 lb/hr of 2-propanol is to be extracted with 1000 lb/hr of water in a continuous countercurrent extraction column operated at 1 atmosphere and 30°C. The feeds are liquids at 16 psi and 30°C. A1-psi pressure drop across the column is anticipated. The column has four theoretical stages. Use the extraction block to calculate the column's performance. Create a sensitivity study varying the quantity of water used for extraction from 700 to 1800 lb/hr in increments of 100 lb/hr. Create a set of units copied from ENG using °C.

Workshop 11.3b   Use a design specification to determine the quantity of water needed to recover 96% of the alcohol.

Workshop 11.3c   Use the decanter block to develop a four-stage batch extraction solution to Workshop 11.3b. Divide the amount of water calculated in Workshop 11.3b into four equal quantities so that each can be added to the batch four times. Compare the quantity of alcohol extracted to the solution for the continuous case. What do you conclude?

Workshop Notes

Workshop 11.1a through 11.1d   All results are similar because this system is nearly ideal. Workshop 11.1c has an error message relating to the pressure of the feed stream. Note that the feed pressure, 14.696, is lower than the feed stage pressure. Workshop 11.1d's feed is set at 16 psi to fix the problem.

Workshop 11.2b   Results at reflux ratio of 1.2 with a bottoms takeoff of 50 lbmol/hr are similar to Workshop 11.2a. Figure 11.32 shows a plot of liquid composition as a

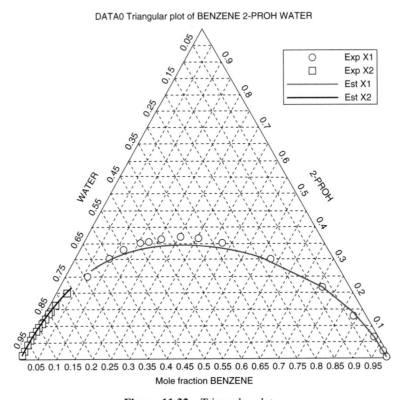

Figure 11.32 Triangular plot.

function of the stage numbers. One may observe that the methanol composition barely changes between stages 6 and 11, which suggests that there are too many stages in the current design.

Workshop 11.2c The column is reconfigured to 11 stages with the feed on stage 7. The results are very close to those of Workshop 11.2b. The column design as proposed by DSTWU has too many stages.

Workshop 11.3a The triangular plot is shown in Figure 11.32.

Workshop 11.3c To obtain the total quantity of 2-propanol extracted, it is convenient to set up a Mix block and feed it the aqueous phases from each decanter. When this is first run, one observes that with one-fourth of the required water added to the the organic feed, no phase splitting occurs. It is necessary to feed both products of the first decanter to the second decanter, and then everything proceeds as expected. If an error message referring to insufficient iterations arises, double the default value of 30.

When comparing the results to Workshop 11.3b, one observes that less than half of the alcohol is extracted by batch extraction.

REFERENCES

Aspen Plus version 7.0, Help documentation, Appendix A.

Boston, J. F. and Britt, H. I., *Comput. Chem. Eng.*, 52, 52–63 (1974).

Boston, J. F., and Sullivan, S. L. Jr., *Can. J. Chem. Eng.,* 52, 52–63 (1974).

Fenske, M. R., *Ind. Eng. Chem.*, 24, 482–485 (1932).

Gilliland, E. R., Multicomponent rectification, *Ind. Eng. Chem.*, 32, 1220 (1940).

Kirschbaum, E., and Gerstner, F., *Verfahrenstechnik*, 1, 10 (1939).

Liddle, C. J., Improved shortcut method for distillation calculations, *Chem. Eng.*, 75(23), 137 (Oct. 21, 1968).

McCabe, W. L., and Thiele, E. W., *IEC.*, 17, 605–611 (1925).

Molokanov, Y. K., Korablina, T. P., Mazurina, N. I., and Nikiforov, G. A. *Inst. Chem. Eng.*, 12(2), 209–212 (1972).

Naphtali, L. M., and Sandholm, D. P., *AIChE J.*, 17, 148–153 (1971).

Seader, J. D., and Henley, E. J., *Separation Process Principles*, Wiley, Hoboken, NJ, 1998, pp. 569–579.

Wilson, A. and Simons, E., *Ind. Eng. Chem.*, 41, 2214 (1952).

Udovenko, V. V., and Mazanko, T. F., *Zh. Fiz. Khim.*, 38 (1964).

Underwood, A. J. V., *J. Inst. Pet.*, 32, 614–626 (1946).

Winn, F. W., *Pet. Ref.*, 37(5), 216–218 (1958).

CHAPTER TWELVE

PROCESS FLOWSHEET DEVELOPMENT

Prior to this chapter, most of the Aspen Plus blocks have been introduced and the reader has gained experience in their use. This chapter deals with the integration of blocks into a process flow diagram (PFD), which is the source document for the process engineering function of producing process and instrument diagrams (P&IDs). A typical PFD contains process details of each processing unit and all reaction units, connected by the main process lines. Each unit operation and reaction unit will contain basic design data, such as operating temperature and pressure, and limited equipment design data, such as the number of theoretical stages required. Additionally, the presence of trace components will be identified, although it is not unusual that some surprises arise. Very few detail design data are included on a PFD, nor are instrumentation and process control requirements, operating instructions, equipment redundancy, safety considerations, shutdown and startup requirements, and the myriad other details that are the responsibility of the P&ID developers.

The PFD development process begins with a process description typically based on small-scale chemical experiments and larger-scale minipilot and pilot-plant scale work. Additionally, a collection of physical and thermodynamic property data is developed by experimentation, collected from the literature, and estimated if necessary. These are used to produce a block diagram of the process, which is usually nothing more than a material balance that includes many performance estimates based on laboratory work and, in some cases, guesstimates.

12.1 HEURISTICS

When developing a process flow diagram, the following experience-based ideas are worthwhile:

Teach Yourself the Basics of Aspen Plus™   By Ralph Schefflan
Copyright © 2011 John Wiley & Sons, Inc.

1. A large flowsheet should be divided into a subset of smaller flowsheets.
2. Each subset should have its own appropriate properties.
3. Initially, in solving a subset, basic blocks without energy balances should be employed and converged.
4. All initial attempts to converge should use Aspen Plus defaults (i.e., Wegstein's method). The Wegstein approach can solve a wide variety of problems.
5. Initial convergence should be with loose specifications.
6. As the development proceeds, the basic blocks should be replaced, one or two at a time, with rigorous blocks, and energy balances turned on.
7. The rigorous blocks should be tested on a stand-alone basis with feeds from the streams calculated using the basic blocks prior to integration.
8. When a subset is executed using rigorous blocks, the independent variables from all tear streams that are feeds to the blocks should be transferred from the converged initial flowsheet where the basic blocks were used.
9. If the Aspen Plus–selected tear streams within the subsets are not suitable, new tear streams should be identified. Convergence blocks should be defined for the new tear streams. New calculation sequences may also be required.
10. When the subsets are completed, assembly of a fully rigorous flowsheet should proceed one subset at a time with intermediate testing. Tear streams may need to be changed. A gradual tightening of specifications is required. This process continues until the combined subsets have been integrated and the entire flowsheet has converged.

12.2   EXAMPLE: THE PRODUCTION OF STYRENE

A flowsheet of a simplified version of the commercial process for the production of styrene by catalytic dehydrogenation of ethylbenzene is shown in Figure 12.1. The process description is as follows.

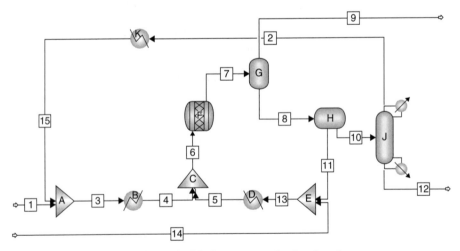

Figure 12.1   Modified styrene production flowsheet.

1. Ethylbenzene is converted to styrene by the catalytic dehydrogenation reaction

$$C_8H_{10}(g) \rightarrow C_8H_8(g) + H_2(g)$$

   The presence of steam in the reactor is known to suppress side reactions.

2. Fresh ethylbenzene, stream 1, and a recycle, stream 15, containing primarily liquid ethylbenzene, combine at A, producing stream 3, which is heated to 500°C at B, producing stream 4.

3. A recycle, stream 11, which is primarily water, is combined with stream 14, a freshwater makeup, at E, producing stream 13 at 50°C.

4. Stream 13 is heated to 700°C, producing stream 5, which is combined with stream 4 at C, producing stream 6 at 560°C.

5. Stream 6 feeds the reactor, F. Stream 7 exits the reactor at a temperature and pressure of 560°C and 14.7 psi, respectively. The reaction conversion is 35%.

6. Stream 7 is cooled to 50°C at G, producing stream 9, which is primarily H_2, and goes to another part of the plant. Stream 8 is cooled further to 25°C at H, to separate the water, stream 11, and organics, stream 10.

7. Ethylbenzene and styrene in stream 10 are separated at J, producing the product stream 12, containing primarily styrene, and a stream containing primarily ethylbenzene, which is condensed at K, producing stream 15.

The process model is to be based on the following:

Stream 1: 100 lbmol/hr of pure ethylbenzene at 25°C and 14.7 psi
Stream 14: 40 lbmol/hr of pure water at 25°C and 14.7 psi

The entire process operates at approximately 1 atmosphere.

12.3 A MODEL WITH BASIC BLOCKS

The first model of this process was developed using split blocks for the separations. The process description above was used to estimate the split values and a simple conversion based block was used to describe the reaction. The model results are given at Chapter Twelve Examples/Example Twelve-1. Analysis of the results seem to agree with the process description.

12.4 PROPERTIES

There are three separations that require rigorous modeling in the development of a process flow diagram. All the necessary pure component data are available in the Aspen Plus data bank. Vapor–liquid, equilibrium and liquid–liquid equilibrium data are required to model the flash, the decantation, and the distillation elements of the process. The Uniquac binary parameters stored in Aspen Plus will be used as a source for the equilibrium data. Property analysis runs designed to present vapor–liquid equilibria for the three possible systems exclusive of hydrogen are shown at Examples/ExampleTwelve1-a. Figure 12.2 shows the binary parameter data

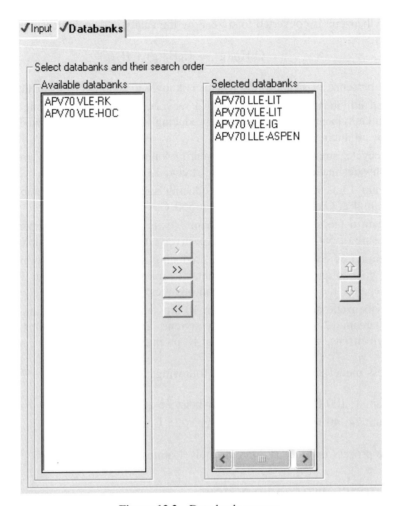

Figure 12.2 Data bank sources.

source choices for the systems, Figure 12.3 shows the Uniquac binary parameters for the three systems, and Figure 12.4 shows a plot of the ethylbenzene–styrene vapor–liquid equilibrium and hand calculations indicate relative volatilities between 1.2 and 1.4. The other two systems exhibit mutual solubility in a very narrow range of composition, and because of this, only liquid–liquid equilibrium is available. These values should be suitable for use in the distillation calculations.

12.5 RIGOROUS FLASH AND DECANTER

A more rigorous version of the process may be found at Examples/ExampleTwelve-2. A two-phase flash and a decanter block were placed onto the flowsheet replacing the splitters at G and H, using the properties from above. Several exploratory runs were made, with the following results. With the two-phase flash at G at $50°C$, the water in the process was pushed out at product stream 9, along with the hydrogen. Subsequently, the flash temperature was changed to $15°C$.

Input	✔Databanks			

Parameter: UNIQ   Data set: 1

Temperature-dependent binary parameters

Component i	C8H8 ▼	C8H8 ▼	EBZ ▼
Component j	EBZ ▼	WATER ▼	WATER ▼
Temperature units	C	C	C
Source	APV70 VLE-LIT	APV70 LLE-LIT	APV70 LLE-LIT
AIJ	0.0	0.0	0.0
AJI	0.0	0.0	0.0
BIJ	-239.3595000	-889.4500000	-968.3700000
BJI	173.3769000	-331.6500000	-354.2300000
CIJ	0.0	0.0	0.0
CJI	0.0	0.0	0.0
DIJ	0.0	0.0	0.0
DJI	0.0	0.0	0.0
TLOWER	90.00000000	20.00000000	20.00000000
TUPPER	100.0000000	40.00000000	40.00000000
EIJ	0.0	0.0	0.0
EJI	0.0	0.0	0.0

Figure 12.3   Uniquac binary parameters.

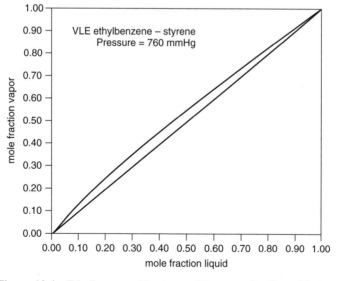

Figure 12.4   Ethylbenzene (1)–styrene (2) vapor–liquid equilibrium.

Summary | Define Variable |

Row / Case	Status	VARY 1 14 MIXED TOTAL MO LEFLOW LBMOL/HR	H2PROD	ORGPRD LBMOL/HR	FLOS6 LBMOL/HR	XH2OS6	XH2OS9
1	OK	3	0.97104011	0.69444454	284.574025	0.01155593	0.02290706
2	OK	5	0.9520221	0.69436453	287.556289	0.02188353	0.04194280
3	OK	7	0.93373108	0.6942888	290.74321	0.03267869	0.06025267
4	OK	9	0.91613065	0.69421855	294.244026	0.04425811	0.07787347
5	OK	11	0.89918026	0.69415586	298.294932	0.05730372	0.09484642
6	OK	13	0.88283443	0.69410549	303.592594	0.07381347	0.11121841
7	Errors	15	0.87323275	0.69419504	327.277253	0.14085932	0.120824

Figure 12.5   Sensitivity study varying the water feed.

A sensitivity study was made varying the amount of water entering the process for the purpose of determining the effect on the products and the composition and flow rate of the reactor feed. The water feed rate was varied from 3 to 15 lbmol/hr. The results are shown in Figure 12.5. The variables are defined below.

- H2PROD is the mole fraction hydrogen in stream 9.
- ORGPRD is the flow rate of water in lbmol/hr in stream 12.
- FLOS6 is the flow rate of stream 6 in lbmol/hr.
- XH2OS6 is the mole fraction of water in stream 6.
- XH2OS9 is the mole fraction of water in stream 9.

After about 13 lbmol/hr of water fed per 100 lbmol/hr of ethylbenzene, the material balance is unstable.

1. The mole fraction of water in stream 6, the reactor feed (variable XH2OS6), varies from about 0.011 to 0.074.
2. The mole fraction of water in stream 9, the hydrogen product (variable XH2OS9), varies from 0.023 to 0.11, which corresponds to a change in the composition of hydrogen (variable H2PROD), varying from 0.97 to 0.88.
3. The total reactor feed flow rate (variable FLOS6) varies from 284 to 327 lbmol/hr.
4. The flow rate of the organic product varies very little; however, rigorous distillation has not yet been implemented.

The question of how much water to use in the process must be reviewed in light of the foregoing results. Assuming that it is desired to produce as pure a hydrogen product as possible, the remainder of the analysis will use a 3-lbmol/hr feed of water, which corresponds to about 1% water in the reactor feed. The process description did not indicate what quantity of water is required in the reactor. Thus, 40 lbmol of water per 100 lbmol of feed is not suitable with the process description given.

12.6 ANALYZING THE RIGOROUS DISTILLATION

The distillation feed from Example 12.2 using a 3-lbmol/hr water feed was employed for the preliminary calculations in preparation for designing and then simulating the column. The feed with appropriate column specifications was employed with a DSTWU block to obtain an estimate of the number of stages, feed stage location, and reflux ratio required. These results may be found at Examples/ExampleTwelve-3a. The results were used to explore the design parameters of the column with several RadFrac runs. The results of these runs culminated in a Sensitivity study with Radfrac at Examples/ExampleTwelve-3b. The desired styrene product composition was about 0.98 mole fraction. Figure 12.6 shows the effect of reflux ratio on the two product compositions. Here S12XE and S2XS are the mole fractions of ethylbenzene in stream 12 and styrene in stream 2, respectively. The column operates with 60 total stages, feed on stage 30, and a distillate rate of 183 lbmol/hr. The external reflux ratio required is about 9. Note that there is a slight amount of hydrogen in the feed; therefore, the use of a partial condenser is imperative.

12.7 INTEGRATING THE RIGOROUS DISTILLATION INTO THE FLOWSHEET

Introducing the converged rigorous distillation results of ExampleTwelve-3b into the converged flowsheet of ExampleTwelve-2 with the sensitivity study removed produces Example Twelve-4a. This example does not converge and produces several errors,

Row / Case	Status	VARY 1 B1 COL-SPEC MOLE-RR	S12XE	S2XS
1	OK	6	0.03366703	0.02490138
2	OK	7	0.02211972	0.01869718
3	OK	8	0.01544934	0.01511328
4	OK	9	0.01126580	0.01286552
5	OK	10	0.00858002	0.01142249

Figure 12.6 Effect of reflux ratio on product composition.

```
   Block: B          Model: HEATER

   Block: C          Model: MIXER

> Loop $OLVER01 Method: WEGSTEIN      Iteration     5
   1 vars not converged, Max Err/Tol   -0.11260E+02

   Block: F          Model: RSTOIC

   Block: G          Model: FLASH2

   Block: H          Model: DECANTER

   Block: E          Model: MIXER

   Block: D          Model: HEATER

   Block: J          Model: RADFRAC
 *** SEVERE ERROR
     COLUMN NOT IN MASS BALANCE.
     CHECK FEEDS, PRODUCTS, AND COL-SPECS SKWS.

 *    WARNING
      REST OF BLOCK BYPASSED DUE TO SEVERE ERROR.


   Block: K          Model: HEATER
 *    WARNING
      ZERO FEED TO THE BLOCK.  BLOCK BYPASSED

   Block: A          Model: MIXER

   Block: B          Model: HEATER

   Block: C          Model: MIXER

> Loop $OLVER01 Method: WEGSTEIN      Iteration     6
# Converged                  Max Err/Tol    0.50527E+00

->Generating block results ...

   Block: K          Model: HEATER

->Simulation calculations completed ...
```

Figure 12.7   Aspen Plus run control panel.

some of which are displayed in the run control panel, a part of which is shown in Figure 12.7. The execution history shows that the RadFrac block and the Heater block that follows it have been bypassed. It would be wise to refer to the process flowsheet shown in Figure 12.8, where the failed blocks are clearly identified. What could be misleading is the message at the end of the execution history, which states that the loop $OLVER01 has converged. A look at the feed to the RadFrac block shows that it is significantly different from that of ExampleTwelve-3b. It is important that the remainder of the flowsheet produce the proper feed. The difficulty is that the sequence of calculations and the initial estimate of zeros for the variables of stream 6, the tear stream chosen by Aspen Plus, do not assure that the feed to the distillation column

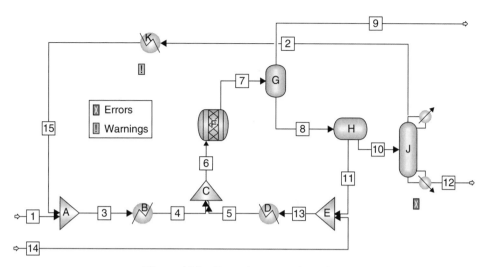

Figure 12.8   Execution errors identified.

from ExampleTwelve-3b enters the column. Because of this it will be necessary to alter the sequence of calculations and provide good estimates for the new tear stream.

12.7.1   Selection of a Tear Stream

If the process flow diagram (Figure 12.8) is examined, one can see that guessing the initial values of either stream 6 or stream 8 permits the sequential modular calculations to pass through each block, finally arriving at the stream values guessed; however, using stream 8 assures that only the organics pass through to the distillation column. One may then copy the converged values from Example Twelve-3b stream 8 onto the appropriate input of Example Twelve-4b. The definition of the tear stream, however, requires the definition of a convergence block for stream 8. This is accomplished by opening the Convergence entry under the list of specifications, opening the Convergence subentry, and selecting New with the object manager. Figure 12.9 shows the selection of tear stream with its convergence block.

12.7.2   Sequence of Calculations

The list of specifications also shows the entry Sequence, which when selected opens the object manager to enable the creation of a sequence of calculations by selecting New. The new execution sequence is shown in Figure 12.10. When this is executed, a satisfactory solution is obtained for each operation. Results are given at Examples/Example Twelve-4b.

12.8   REACTOR FEED

The process description, item 4, states: "Stream 13 is heated to 700°C, producing stream 5, which is combined with stream 4 at C, producing stream 6 at 560°C". In

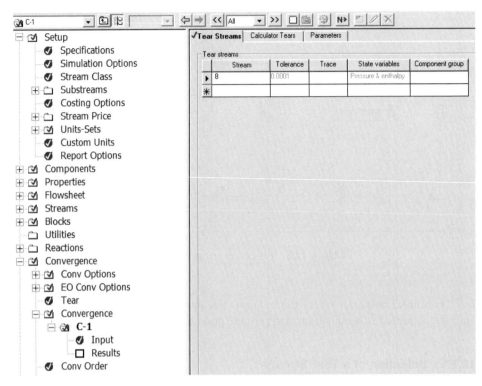

Figure 12.9 Tear stream selection.

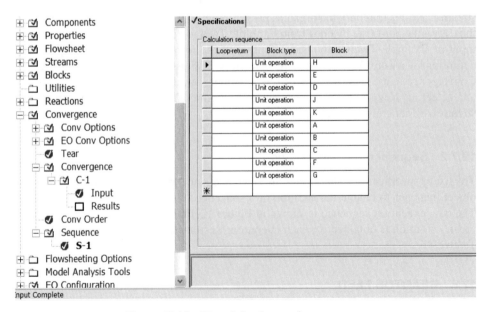

Figure 12.10 User-defined execution sequence.

actuality it is necessary merely to provide 560°C as the set point to heaters B and D. This is implemented at Example Twelve-4b.

12.9 MISCELLANEOUS CONSIDERATIONS

Pumps are not used at this level of development. It is understood that the use of a mostly constant 14.7 psi pressure in this flowsheet is technically incorrect but introduces only a small error. The feed to the distillation column was changed to 18 psi to accommodate the column's internal pressure. The actual size and location of the equipment has not been defined since the physical layout there has not yet been developed; therefore, pump selection cannot proceed since the actual service is unknown. For example, the distillation column has 60 theoretical stages, and if the efficiency is 100% and the tray spacing is 2 ft, the column is at least 120 ft tall (a 12-story building). If the site is in a windy area, this column may be built as two 60-ft columns with pumps in between.

Similarly, the type of equipment used for heating some streams to high temperatures and cooling off others is unknown; thus, detailed design will be carried out by others.

The amount of water used to control the formation of by-products is unknown. The reaction kinetics is unknown, and therefore at this point in the flowsheet development, a reaction conversion of 35% is used. Scale-up would be from laboratory data.

12.10 WORKSHOPS

Workshop 12.1 Figure 12.11 shows a section of a process for converting toluene to benzene in a hydroalkylation reactor. The main reaction is

$$C_7H_8 + H_2 \rightarrow C_6H_6 + CH_4$$

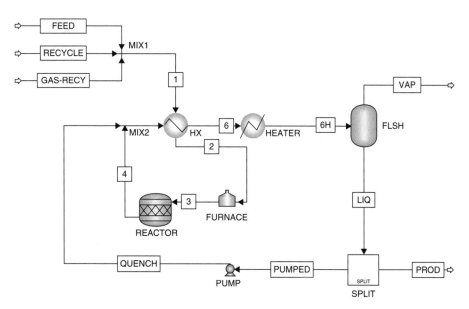

Figure 12.11 Process for production of benzene.

TABLE 12.1 Process Feeds

Component	Feed (lbmol/hr)	Recycle (lbmol/hr)	Gas Recycle (lbmol/hr)
Hydrogen			2045.9
Methane			3020.8
Benzene		3.4	42.8
Toluene	274.2	82.5	5.3
Biphenyl		1.0	
Temperature ($^\circ$F)	75	250	121
Pressure (psi)	569	569	569

TABLE 12.2 Process Operating Conditions

Location	Pressure (psi)	Temperature ($^\circ$F)
Furnace inlet	569	1000
Furnace outlet	494	1200
Reactor outlet	494	1268
Pump outlet	494	
Heater outlet	480	180
Flash	480	

An unavoidable side reaction that produces biphenyl is

$$2C_6H_6 \rightarrow C_{12}H_{10} + H_2$$

The conditions for the feed and two recycle streams are given in Table 12.1. The specifications in Table 12.2 are the pressures and temperatures at various points in the process.

The flow rate of the quench stream is to be such that the temperature of the combination of reactor effluent and the quench stream should be 1150°F. Conversion of toluene is 75%. Two percent of benzene present after the first reaction occurs is converted to biphenyl. The overall heat transfer coefficient in the heat exchanger is 60 Btu/(hr-ft^2-$^\circ$F). Develop an approximate material and energy balance for the process.

Workshop 12.2a Figure 12.12 shows a process to create a mixture suitable for use as a natural gas and recover a light hydrocarbon liquid from the feed given in Table 12.3. The feed is at 80°F and 237.7 psi. Develop a first estimate material and energy balance for this process. The equipment is to be modeled simply. The specifications for the processing equipment are as follows:

Chiller: outlet at −40°F and 230 psi.

Gas/gas heat exchanger: modeled as two heaters with a heat stream used to pass heat between them.

Cold-side heater: outlet at 160 psi. A 4°F approach at the cold-stream outlet of the heat exchanger (i.e., the temperature of the inlet of the hot-side heater minus the temperature of the outlet of the cold-side heater).

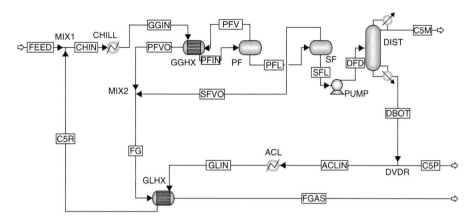

Figure 12.12 Process for production of natural gas.

TABLE 12.3 Component List

ID	Formula	Databank Name	Lb-mol/hr
N2	N_2	Nitrogen	52.0
CO2	CO_2	Carbon dioxide	89.12
C1	CH_4	Methane	6033.50
C2	C_2H_6	Ethane	297.08
C3	C_3H_8	Propane	141.62
IC4	$3(CH_3)CH$	Isobutane	31.84
NC4	$2(CH_3)2(CH_2)$	n-Butane	45.11
IC5	$(2CH_3)CHCH_2CH_3$	2-Methylbutane	17.60
NC5	$CH_3 \cdot 3(CH_2)CH_3$	n-Pentane	14.90
NC6	$CH_34 \cdot (CH_2)CH_3$	n-Hexane	19.24

Hot-side heater: outlet at 225 psi.

Primary flash: adiabatic at 160 psi.

Secondary flash: adiabatic at 100 psi.

Pump: outlet at 260 psi.

Distillation column: use DSTWU with NC4 and IC5 as the light and heavy key components, respectively; 99% of the light key is to be in the distillate and 99% of the heavy key is to be in the bottoms of the column. A partial condenser is required. The top of the column is at 240 psi and the bottom at 255 psi.

Recycle splitter: 20% of the column bottoms are to be returned to the process.

Air cooler: outlet at 100°F and 250 psi.

Gas/liquid heat exchanger: modeled as two heaters with a heat stream used to pass heat between them.

Hot-side heater: outlet at 80°F and 237.7 psi.

Cold-side heater: outlet at 100 psi.

A partial setup for this flowsheet exists at Workshops/Twelve-2. It is complete except for implementation of the specifications for the gas/gas heat exchanger.

Complete this implementation and run the workshop. The results are given at Workshops/Twelve-2a.

Workshop 12.2b   Replace all of the heaters in Workshop 12.2a with countercurrent heat exchangers, using the shortcut method to implement the design specifications given above. Run Aspen Plus to determine the tear streams. Compare the number of tear streams to those in Workshop 12.2a. Where necessary, substitute the appropriate values obtained from Workshop 12.2a to aid in converging the model. The results are given at Workshops/Twelve-2b.

Workshop 12.2c   Convert the heat exchanger design from above to shortcut simulation models and rerun the simulation. The results are given at Workshops/Twelve-2c.

Workshop 12.2d   Design the distillation column using RadFrac at twice the minimum reflux ratio and a distillate rate specification such that the column meets the specifications above. Select a suitable design to prepare for using it to replace the existing DSTWU model. The results are given at Workshops/Twelve-2d.

Workshop 12.2e   Replace the DSTWU model with the RadFrac model from Workshop 12.2d. Make the necessary Aspen Plus run(s) required to develop a complete material and energy balance. If necessary, substitute the appropriate values obtained from Twelve-2a to aid in converging the model. The results are given at the Workshops/Twelve-2e.

Workshop Notes

Workshop 12.2a   The placement of a pump is appropriate because of the large changes in pressure.

Workshop 12.2b   The shortcut design method does not provide any design details but merely estimates the area required. It uses a constant overall heat transfer coefficient with a default of about 150 Btu/(hr-ft^2-$^\circ$F). The best of poor choices would be to obtain an estimate of U from a literature source such as Perry and Green (1999). A proper design would involve use of the EDR models referred to in Chapter Nine.

Workshop 12.2c   Conversion of the design to a rating model consists merely of providing the area calculated in 12.11. Convergence of a small exchanger will probably require the specification of a smaller area convergence tolerance than the default value.

Workshop 12.2d   Using the DSTWU as an approximate solution, the RadFrac model solution space was searched with a few trial runs settling on about 35 trays, feed onstage 18, and a reflux ratio of about 1.3. For a duplicator block fed a DSTWU (for comparison) and a RadFrac block with the same feed, after settling on a D/F value of 0.641, a sensitivity study varying the reflux ratio was made. Results show that a reflux ratio of 1.2 meets the recovery specifications.

Workshop 12.2e   Caution on defining the overhead product state of the distillation column. Try using a partial condenser and observe that the process converges and the results are what one might expect. Change to a total condenser and observe that the process converges. But is it correct, and does it meet process requirements?

REFERENCE

Perry, R. H., and Green, D. W., *Perry's Chemical Engineers' Handbook on CD-ROM,* 7th ed., McGraw-Hill, New York, 1999, Table 11.2.

CHAPTER THIRTEEN

OPTIMIZATION

Aspen Plus's focus in the area of process optimization is limited primarily to applications involving process models constructed using Aspen Plus. As examples, process optimization can refer to selection of the optimal operating conditions of process equipment, or optimal design of equipment, each subject to the definition of "optimal." These conditions, or design variables, could be driven by "lowest cost," or "easiest to control," or any other basis of evaluation. The applications are typically characterized by a function that is to be maximized or minimized subject to variables that are meaningful within limits; for example, a negative mole fraction has no meaning within a process. In the classical sense, such problems are formally described by an objective function and constraints on the independent variables. A formal example of an objective function Φ is given by

$$\Phi = \Phi(x_1, x_2, \ldots, x_n) \qquad (13.1)$$

where $x_1, x_2, \ldots, x_n$ refer to the independent variables. An example of an equality constraint, f_1, is given by

$$f_1 = f_1(x_1, x_2, \ldots, x_n) = 0 \qquad (13.2)$$

and an inequality constraint, f_2, by

$$f_2 = f_2(x_1, x_2, \ldots, x_n) \leq 0 \qquad (13.3)$$

The variables that define the constraints and the objective function need not necessarily be continuous but may take on only integer values. The objective function has one value for a given set of independent variables. The problem is to find the set of

Teach Yourself the Basics of Aspen Plus™   By Ralph Schefflan
Copyright © 2011 John Wiley & Sons, Inc.

independent variables, subject to the applicable constraints, that yield the optimum value of the objective function. In many circumstances there may be several solutions to an optimization problem.

There are many algorithms for solving linear and nonlinear optimization problems. Good summaries may be found in Beveridge and Schechter (1970) and in Edgar and Himmelblau (2001). Aspen Plus provides the following proprietary methods:

- SQP: successive quadratic programming
- Complex: a black-box pattern search

Each is described in Aspen Plus's Help.

13.1 OPTIMIZATION EXAMPLE

A simple example is the trade-off between the number of theoretical stages and the reflux ratio in a distillation column required to obtain a target product composition. For this case the independent variables are the number of theoretical stages, a discrete variable that takes on integer values, and the reflux ratio, a continuous variable that is constrained between a minimum value which corresponds to a practical value near an infinite number of theoretical stages, and a maximum which corresponds to a minimum number of stages. The objective function has contributions from the following factors.

- Number of theoretical stages, N
- Installed cost of the column, where each stage adds an increment of height to the column, I$ (dollars/theoretical stage)
- Expected life of the column, L (years)
- Annual maintenance of the column, M$ (dollars/year)
- Annual reboiler load, R (Btu)
- Annual condenser load, C (Btu)
- Cost of heating, H$ (dollars/Btu)
- Cost of cooling, C$ (dollars/Btu)
- Operating labor, O$ (dollars/year)

A simplified objective function, Φ, which represents the total cost of the column over its life is defined by

$$\Phi = N^*I\$ + M\$^*L + R^*H\$^*L + C^*C\$^*L + O\$^*L \qquad (13.4)$$

The implementation of the objective function is carried out using the entry Model Analysis, Optimization. The initial input form on which the objective function is defined is given in Figure 13.1. In this example, variables defined in the figure are taken from the process during each iteration, and are used in calculation of the objective function. The various elements that are required for optimization are given in the tabs across the top of the figure. Figure 13.2 provides for the formal definition of the objective function and its constraints. Note that it is possible to code the objective function in the objective function frame in Fortran. Check boxes to select to maximize or minimize

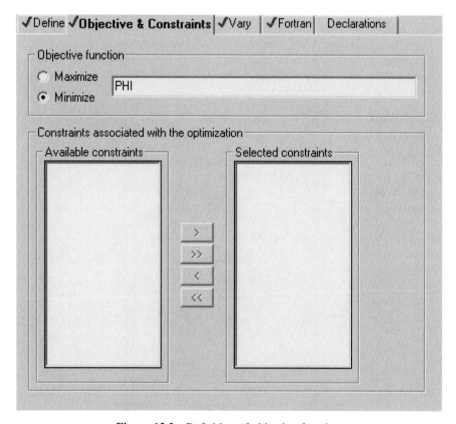

	Flowsheet variable	Definition
	ENSTGS	Block-Var Block=DSTW Variable=ACT-STAGES Sentence=RESULTS
	RDUTY	Block-Var Block=DSTW Variable=REB-DUTY Sentence=RESULTS Units=Btu/hr
	CDUTY	Block-Var Block=DSTW Variable=COND-DUTY Sentence=RESULTS Units=Btu/hr
✳		

Figure 13.1   Definition of variables.

Figure 13.2   Definition of objective function.

the objective function are also available. Alternatively, if the tab Fortran is selected, a box appears, for entering the objective function as a series of statements in Fortran, as shown in Figure 13.3. Figure 13.4 provides the mechanism to vary an independent variable, in this case the reflux ratio, and to define the limits that apply to the variable.

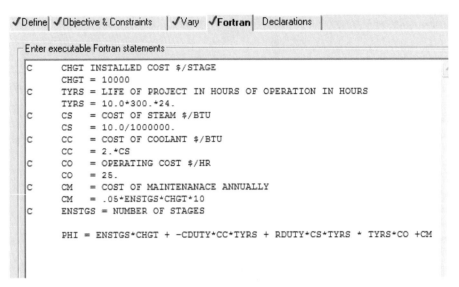

Figure 13.3   Fortran coding of objective function.

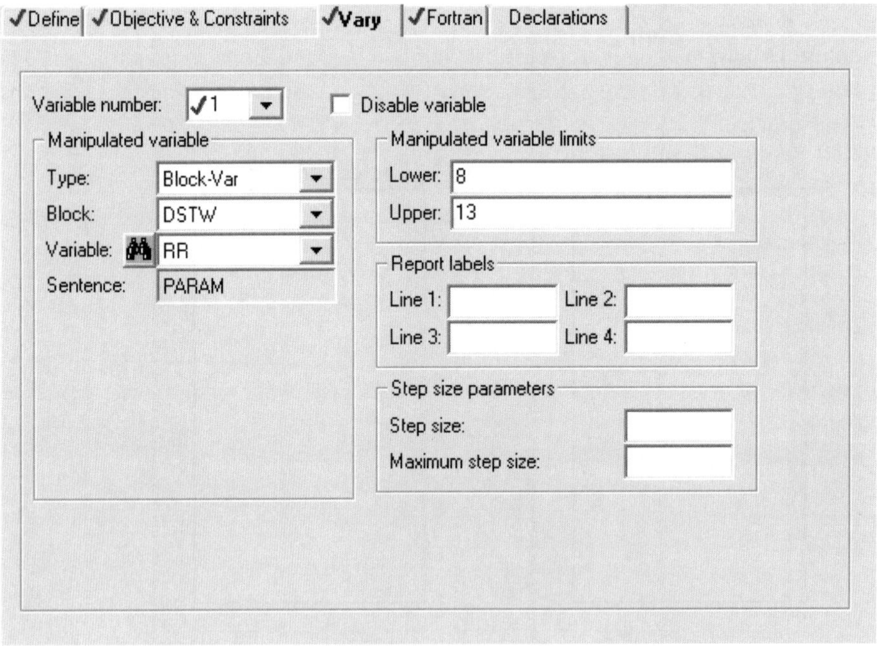

Figure 13.4   Setup to vary independent variable.

Having defined the objective function, it is necessary to develop a model of the process, which, given the feed, the desired product composition, the reflux ratio, the feed location, and the number of theoretical stages, will calculate the reboiler and condenser duties. As an example, the DSTWU block will be used to create the model. A characteristic of the block is that the number of stages is a continuous variable. If a constraint is required, the folder Constraints is selected and variables are selected in

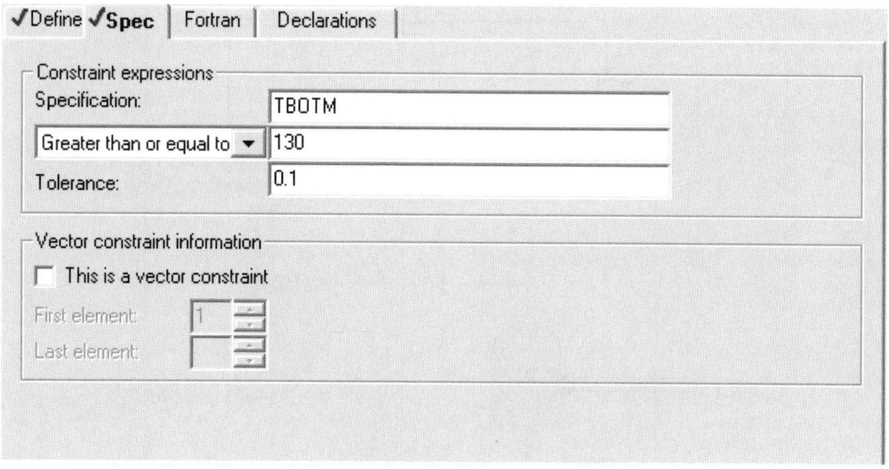

| √Define √**Spec** | Fortran | Declarations |

Constraint expressions
Specification: | TBOTM

Greater than or equal to ▼ | 130

Tolerance: | 0.1

Vector constraint information
☐ This is a vector constraint
First element: | 1
Last element: |

Figure 13.5 Constraint.

a display similar to Figure 13.1. Although this example does not require a constraint, an example of the formulation of a constraint is shown in Figure 13.5. Results of the example above are given at Chapter Thirteen Examples/ExampleOne. In this case the nature of DSTWU does not permit simultaneous variation of the reflux ratio and the number of stages required; however, for the case of varying the reflux ratio only, the results show that a tall column with a relatively low reflux ratio produces the lowest value of the objective function.

Although results are available at the Optimization block, additional results at Convergence/Convergence under the tab Iterations show values of the objective function and the independent variable through the solution process.

13.2 WORKSHOPS

Workshop 13.1   Propane at 14.7 psi and 200°F is to be compressed to 120 psi with two single-stage polytropic compressors, using the ASME method, separated by a cooler that reduces the first compressor's outlet temperature to 200°F. Develop an optimization that will select the outlet pressure of the first compressor such that the work required for the system is minimized. The polytropic efficiency for both compressors is 0.72 and their mechanic efficiency is 1.0. The minimum allowable work for both compressors is 10 hp.

Workshop 13.2a   The process shown in Figure 13.6 is to be optimized with respect to the pressure of the flash, the column internal reflux ratio, and its distillate to feed ratio. In preparation it is necessary to develop a series of process models in steps. A key element of the process is the requirement that the column vapor product be compressed to the operating pressure of the flash, which for model development is 150 psi. The flash is to operate adiabatically. It is necessary that a Fortran statement in a calculator block be written to accomplish this. Once the code has been written, the sequence tab is selected and the appropriate option to be employed upon execution is configured. The column has 10 theoretical stages with the feed on stage 5. The column

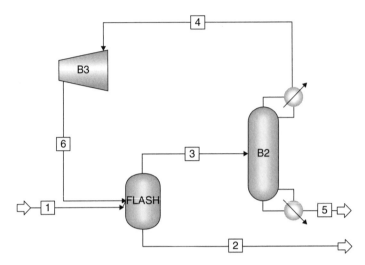

Figure 13.6 Workshop 13.2a process.

TABLE 13.1 Process Feed

Component	Flow Rate (lbmol/hr)
Propane, C3	10
n-Butane, NC4	20
n-Pentane, NC5	20
n-Hexane, NC6	10
Cyclohexane, CC6	40

reflux ratio is 1.0 and the distillate/feed ratio is 0.5. The overhead product is saturated vapor at a pressure of 50 psi, and the column bottoms pressure is 51 psi. This model should not incorporate optimization. The feed to the process is a saturated liquid at 250 psi and its composition is given in Table 13.1.

Workshop 13.2b Develop an optimization, maximizing profitability, of the flowsheet above by varying the flash pressure between 150 and 250 psi. The objective function needs to be coded in Fortran and will incorporate the following terms:

Stream 5 n-hexane value: $0.1/lbmole
Stream 5 cyclohexane value: $0.20/mole
Coolant cost: 1.0×10^{-6}/Btu
Heating media cost: 1.0×10^{-7}/Btu
Net work cost: $0.020/hp

Workshop 13.2c Develop an optimization of the flowsheet in three dependent variables. In addition to flash pressure, the column distillate/feed ratio varies between 0.4 and 0.7. The internal reflux ratio varies between 0.4 and 0.7 using the same objective function as above.

Workshop Notes

Workshop 13.1   A Calculator block is required to pass the outlet pressure of the compressor to the pressure specification of the Heater block. Constraints on both compressor horsepower limitations have to be set up and selected on the Objective Function and Constraints tab of the Optimization display.

Workshop 13.2a   The Calculator setup is given in Figures 13.7, 13.8, and 13.9.

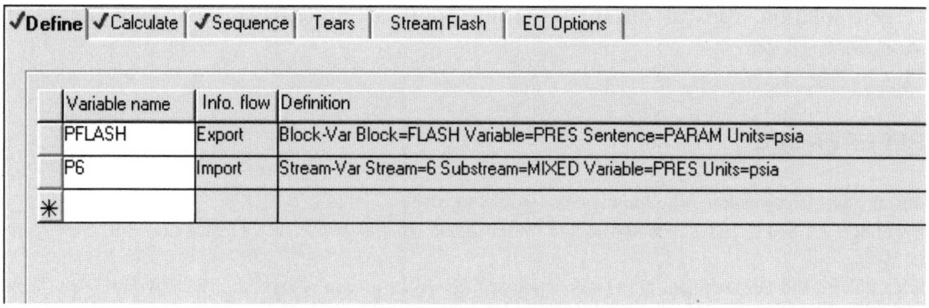

Figure 13.7   Defining calculator variables.

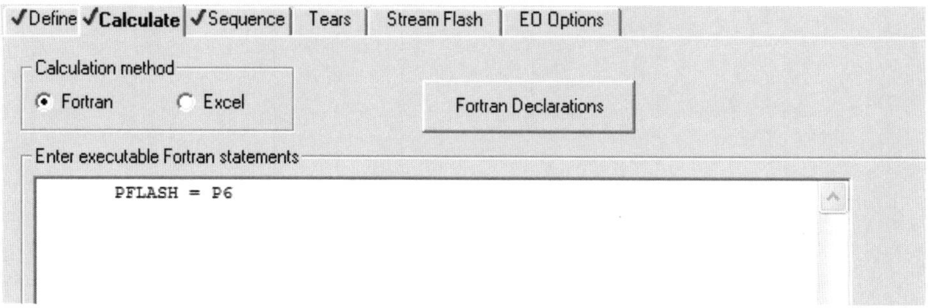

Figure 13.8   Defining calculator calculations.

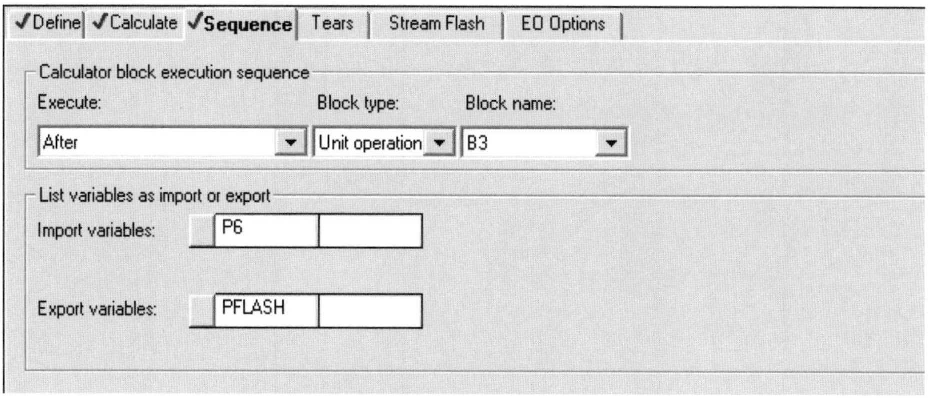

Figure 13.9   Defining calculator sequence.

Workshop 13.2b   To facilitate the passing of pressure from the compressor to the flash, it is necessary to set the flash pressure to zero on its input form. This problem was run several times at starting compressor pressures, varying from 250 to 150. In all cases the SQP algorithm converged, sometimes with output pass errors, but in all cases a value of the final objective function was displayed under the Convergence tab. Results showed that a pressure approaching 150 psi yielded the lowest value of the objective function. Results of the SQP algorithm need careful checking.

Workshop 13.2c   Although the optimization with three variables gave a converged solution, runs at different starting points exhibited the same behavior as in Workshop 13.2b.

REFERENCES

Aspen Plus Help, version 7.0, Optimization—Convergence.

Beveridge, G. S. G., and Schechter, R. S., *Optimization: Theory and Practice*, McGraw-Hill, New York, 1970.

Edgar, T. F., and Himmelblau, D., *Optimization of Chemical Processes*, 2nd ed., McGraw-Hill, New York, 2001.

CHAPTER FOURTEEN

COMPLEX EQUILIBRIUM STAGE SEPARATIONS

Complex equilibrium stage separations typically involve highly nonideal systems and frequently require a literature search to locate vapor–liquid and/or liquid–liquid equilibrium data when they are not available in Aspen Plus's data bank. Often data are not available in the literature, so that estimation by Unifac is the last resort. In many cases, data at more than one set of conditions are involved: for example, azeotrope composition at two or more pressures. Sometimes it is necessary to employ both vapor–liquid and liquid–liquid equilibrium data, and it may be necessary to define unique property sets for each column, decanter, and so on. These types of separations involve the following general areas:

- Energy-integrated systems
- Homogeneous azeotropic distillation
- Extractive distillation
- Heterogeneous systems

The basic ideas and analytical techniques applicable to the applications above are described in the following sections.

14.1 ENERGY INTEGRATION APPLICATIONS

A common technique is to make use of the energy available in the bottoms product of a distillation column. Since the bottoms temperature is almost always higher than that of the feed, a bottoms-feed heat exchanger is frequently employed. Modeling of this idea is trivial, and a modest amount of energy can be saved.

Teach Yourself the Basics of Aspen Plus™ By Ralph Schefflan
Copyright © 2011 John Wiley & Sons, Inc.

A method that has much greater potential is applicable when two columns operate at temperatures such that the overhead temperature of one column is at a higher temperature than the bottoms of the other, and the net reboiler duty of the first is on the same order of magnitude as the condenser duty of the other. The procedure is to use one side of a single heat exchanger as the reboiler of column 1 and the other side as the condenser for column 2. This is illustrated in Figure 14.1. In this example a small heat exchanger is used to make up for the insufficiency of the net heat flow between heaters 1 and 2.

One may note that both columns' configurations have changed considerably. The first column is reconfigured as a refluxed stripper, and only one degree of freedom, such as the reflux ratio, is available. The bottoms product of the column is split in a ratio that is equivalent to the second degree of freedom if the column were not energy integrated: for example, the value of the vapor/distillate ratio, or the flow leaving the splitter.

The second column is configured as a reboiled absorber, which also has only one degree of freedom. This could be the boil-up ratio or the bottoms product rate. The reboiler–condenser is modeled as a pair of heaters with a heat stream flowing from the condenser half to the reboiler half. The overhead splitter returns reflux after the overhead flow is condensed by the reboiler–condenser heat exchanger.

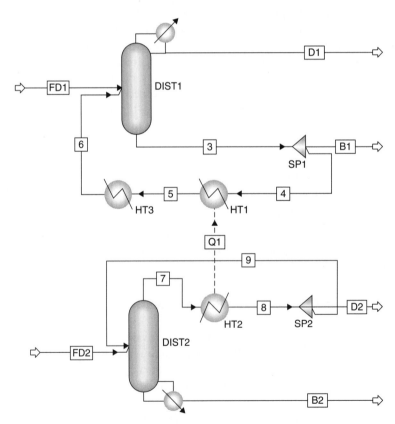

Figure 14.1   Shared reboiler–condenser flowsheet.

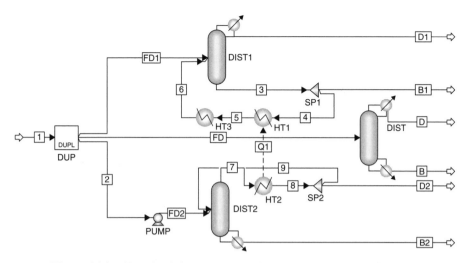

Figure 14.2 Shared reboiler–condenser flowsheet with reference column.

When implementing this approach it would be prudent first to have developed a model of each column without energy integration to facilitate setting up specifications for the degrees of freedom. An example of such a reference column, Dist, is illustrated in Figure 14.2. This example employs a benzene–toluene distillation as its basis, and the demonstration of the energy savings uses two columns each, virtually identical to the reference column. Although the original column, Dist, is not part of the scheme, it is convenient to have it available for debugging and as an aid to setting the specifications for the degrees of freedom. In this example the reboiled absorber operates at a higher pressure than the refluxed stripper, to facilitate heat transfer between the bottoms of the refluxed stripper and the reboiled absorber's distillate. Unifac is used merely as a convenience to provide estimated vapor–liquid equilibrium and is not recommended for a real application. When the model is executed using Aspen Plus's defaults, it does not converge. Initial estimates were provided for both secondary column feeds to facilitate convergence. Details are given at Chapter Fourteen Examples/ExampleOne. The difference in performance in columns Dist and Dist2 is due to the higher pressure in Dist2. The higher pressure reduces the relative volatility of the benzene–toluene system.

The energies used in the reference column (Dist) are as follows:

- Condenser duty: $-2{,}005{,}480$ Btu/hr
- Reboiler duty: $2{,}019{,}100$ Btu/hr
- Heat transferred through stream Q1: $1{,}831{,}371$ Btu/hr
- Reboiler energy use saved (Dist1): 91.3%
- Condenser energy use saved (Dist2): 100%

The reboiler duty will still need to be provided for the source column, Dist2, and the sink column, Dist1, will still require that its condenser duty be provided. This type of configuration for several columns is analogous to a multieffect evaporator.

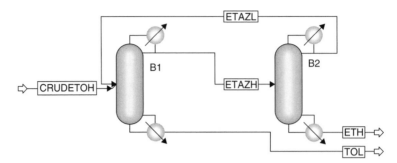

Figure 14.3   Homogeneous azeotropic distillation flowsheet.

14.2   HOMOGENEOUS AZEOTROPIC DISTILLATION

Figure 14.3 shows an example of the configuration of two distillation columns set up for homogeneous azeotropic distillation. The intent of the separation is to produce nearly pure ethanol from a mixture of ethanol and toluene whose composition is below the azeotrope's composition. The arrangement of the two columns takes advantage of the difference in azeotrope composition at different pressures. To facilitate the following example, Aspen Plus's stored Wilson equation binary parameters, which are shown in Table 14.1 are used to estimate the composition of the azeotrope as a function of pressure. If these were to be used for an industrial application, the results would need to be checked against experimental data.

The basic idea of the separation is explained with reference to Figure 14.3. A simple material balance shows that the larger the pressure differential between the two azeotropes, the smaller the recycle flow ETAZL. For illustration purposes, the higher-pressure column, B1, is selected to operate at 760 mmHg and the lower-pressure column, B2, at 50 mmHg. The stream ETAZH's composition is approximately 0.815 mole fraction ethanol and feeds the column B2. The stream ETAZL's composition is approximately 0.685 mole fraction ethanol and is a second feed to column B1. The bottoms product of column B2 is nearly pure ethanol, and the bottoms product of column B1 is nearly pure toluene. Other pressures can surely be selected with an analysis of the trade-off between recycle flow and associated operating costs against the costs associated with purchase and installation of the columns. An initial estimate of the design of the two columns can easily be done using the McCabe–Thiele method on both columns. The low-pressure column can be analyzed with the toluene as the more volatile component.

When such a design is undertaken, the following procedure is recommended:

1. Locate a source of experimental data and fit it to an activity coefficient equation.
2. Determine the azeotrope composition as a function of pressure.

TABLE 14.1   Ethanol–Toluene Azeotrope at Various Pressures

Pressure (mmHg)	50	100	400	760	1420
Mole fraction ethanol	0.685	0.715	0.77	0.815	0.85

3. Perform a material balance for the connected columns to estimate the approximate flows and compositions of the intercolumn streams.

4. Design each column rigorously with loose specifications.

5. Tighten the specifications gradually.

6. Connect the columns, loosening the specifications.

7. If convergence is difficult, provide starting estimates for the intermediate streams.

8. If necessary, change the default convergence method.

9. When convergence is achieved, tighten the specifications gradually.

Examples of some of the steps above are given at Examples/etohtolueneazeos, to calculate azeotrope composition as a function of pressure, ExampleTwoMB the material balance–simple blocks calculation, and ExampleTwo the rigorous solution.

14.3 EXTRACTIVE DISTILLATION

Benzene–cyclohexane is an example of a close-boiling system. Kojima et al.'s (1968) vapor–liquid equilibrium data for this system are shown in Figure 14.4. Note that the vapor–liquid equilibrium curve is almost the same as the $y = x$ line, which makes it virtually impossible to separate using standard distillation methods. Systems such as these can be separated by introducing a third component, the solvent, with a lower volatility than either component, into a distillation column. The solvent must have the

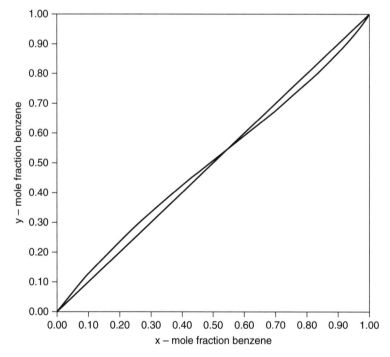

Figure 14.4   Vapor–liquid equilibrium benzene–cyclohexane at 760 mmHg.

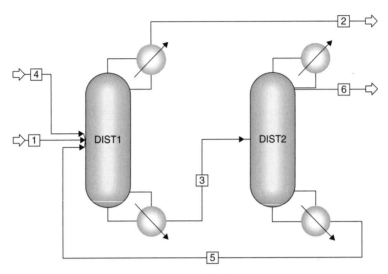

Figure 14.5   Extractive distillation flowsheet.

ability to change the activity coefficients of the binary system such that the relative volatility of the lighter component is increased. Typically, the solvent is introduced near the top of the column and is withdrawn at the bottom along with the heavier component. The solvent is separated from the heavier component in another column, and the solvent is recycled. Such a process is shown in Figure 14.5.

When designing such a process, in addition to determining the number of theoretical stages required and the primary feed location, one must deal with the following issues:

- Solvent selection
- Solvent use per mole of feed
- Stage for introduction of the solvent into the column

There are no hard and fast rules for solvent selection. A list of potential solvents is given by Van Winkle (1967), and general guidelines for solvent selection based on the polarities of the components to be separated are given by Wankat (1988).

The initial step in designing a system is to evaluate a proposed solvent. Does it increase the relative volatility of the lighter component? How much solvent is required? As an example, Seader and Henley (2006) suggest that aniline would be a good solvent for the benzene–cyclohexane system. This can be evaluated by creating a series of $x-y$ diagrams on a solvent-free basis using the ratio of solvent to nonsolvents as a parameter. The calculations to generate such data are given at Examples/ExampleThreeData. These data were transferred to the Excel spreadsheet Examples/benzenecyclohexaneaniline.xls, where the calculations to produce $x-y$ data on an aniline-free basis are given. The results are displayed in Figure 14.6. One may observe that as the ratio of aniline to cylohexane plus benzene increases, there is a dramatic effect on the relative volatility of the cyclohexane–benzene system.

Prior to modeling the complete system it is important that each column's design be undertaken on a stand-alone basis. For the first column, initial calculations to establish the number of stages, reflux ratio, bottom specification, and feed locations are

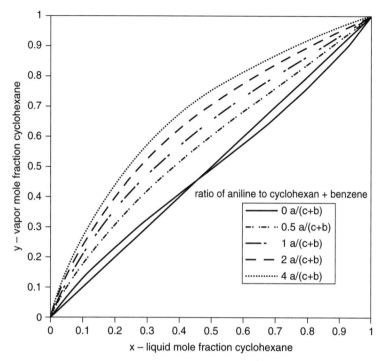

Figure 14.6 Effect of aniline addition on relative volatility in a benzene–cyclohexane system.

incorporated in a sensitivity study to develop the effect of solvent flow rate on product compositions. This study is given at Examples/ExampleThreeS. The results of the study are shown in Figure 14.7. An optimum occurs at a solvent flow rate of 300 lbmol/hr.

The design of the complete system to produce products that are at least 95% pure products is given at Examples/ExampleThree2. It is important to note that the amount of makeup solvent is difficult to determine in advance; therefore, a series of runs ExampleThree1, ExampleThree2, and ExampleThree3 were executed to determine the effect of solvent makeup on the composition of the products. These were transferred to a spreadsheet, with the results shown in Figure 14.8. An approximate optimum appears at a solvent makeup of 0.4 mol/hr per 100 mol of feed.

14.4 HETEROGENEOUS OPERATIONS

These types of separations usually involve a decanter in the process. A system such as that shown in Figure 14.9, production of purified ethanol at a concentration higher than the ethanol–water azeotrope composition, involves the inclusion of a third component, some times called an *entrainer*, which when combined with the azeotrope forms two liquid phases, one rich in ethanol and the other rich in water. The water-rich product of the decanter is fed to the azeotropic column, Dist1, and the entrainer-rich product to the alcohol-producing column, Dist2. The aqueous product issues from the bottom of Dist1 and ethanol is the bottom product of Dist2.

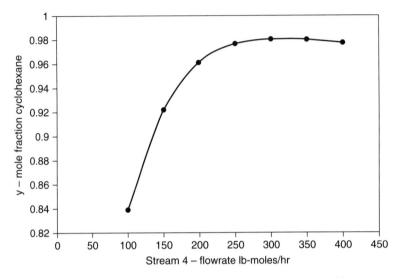

Figure 14.7 Effect of solvent flow rate on product composition.

Stream 4=0.2				Stream 4=0.4				Stream 4=0.6		
streams	2	4		streams	2	4		streams	2	4
	lbmoles/hr	lbmoles/hr			lbmoles/hr	lbmoles/hr			lbmoles/hr	lbmoles/hr
BENZENE	1.6238362	48.243912		BENZENE	0.5219898	49.476921		BENZENE	0.5247517	49.4754588
CYCLOHX	48.0397337	1.9553317		CYCLOHX	49.109847	0.8916487		CYCLOHX	49.107221	0.89254601
ANILINE	0.33643012	0.000756		ANILINE	0.3681634	0.0314307		ANILINE	0.3680273	0.23199515
	mole frac	mole frac			mole frac	mole frac			mole frac	mole frac
BENZENE	0.03247672	0.9610341		BENZENE	0.0104398	0.9816849		BENZENE	0.010495	0.97777587
CYCLOHX	0.96079467	0.0389508		CYCLOHX	0.9821969	0.0176914		CYCLOHX	0.9821444	0.01763924
ANILINE	0.0067286	1.51E-05		ANILINE	0.0073633	0.0006236		ANILINE	0.0073605	0.00458488

Figure 14.8 Effect of solvent makeup flow rate on product composition.

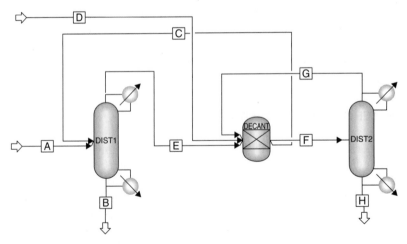

Figure 14.9 Heterogeneous azeotropic distillation flowsheet.

The key to solving such problems is a thorough understanding of the liquid–liquid phase equilibrium. The binary parameter values that determine the decanter performance are critical and must be checked carefully to be sure that the calculations yield the result expected. Excellent sources of liquid–liquid equilibrium data are Arlt et al. (1979–1987) and Stephen and Stephen (1964). When ternary liquid–liquid equilibrium data for a particular component pair is not available, it may be necessary to determine its binary parameters from vapor–liquid equilibrium data.

It is very difficult to merely specify the appropriate blocks in Aspen Plus and expect the process to converge. As an example, an analysis of Figure 14.9 shows that two tear streams are required. The streams about which one potentially knows the most about are the decanter products, streams C and F. It is likely that Aspen Plus's choice of tear streams will need to be changed. The design and simulation of such a system begin with the development of a ternary diagram which shows the various tie lines. The use of a selected tie line and performance specifications with simple blocks permits the development of a material balance. An example is given at Examples/ExampleFour. Starting from this point, rigorous blocks are piecewise inserted into the process, and each of the columns is designed separately. The complete simulation requires considerable refinement, starting with loose specifications and will probably require an increase in the number of theoretical stages required and the reflux ratio as the specifications are tightened. An exact solution to tighter specifications becomes more and more difficult to achieve and should be approached by starting with loose specifications, high reflux ratios, and excessive number of stages.

14.5 WORKSHOPS

Workshop 14.1 A saturated liquid feed of 50 mol % methylcyclohexane, the remainder being *n*-heptane, is to be separated by extractive distillation using phenol as the extracting agent. It is required that the methylcyclohexane product be at least 95 mol % pure and the toluene product be at least 95 mol % pure. The phenol recycle stream is to be at least 99.5 mol % pure. The columns are to operate at 760 mmHg. The binary parameter values stored in the Aspen Plus data bank are adequate. Design the system.

Workshop 14.1a Determine the optimum amount of phenol to be used as an auxiliary feed to the first column, Dist1. Determine the number of theoretical stages required and the location of the feeds.

Workshop 14.1b Place the second column, Dist2, on the process flowsheet of Workshop 14.1a. Use the bottoms product of Dist1 as the feed to Dist2. Design Dist2.

Workshop 14.1c Connect the bottoms of Dist2 to Dist1. Be sure to provide a makeup feed stream to account for the phenol losses from each column. Execute the completed model calculations until convergence is achieved.

Workshop 14.2 A mixture containing 28.1 mol % methanol, 70.9 mol % water, and 1 mol % mesity oxide is to be distilled in a column designed to produce a distillate with at least 99.5 mol % methanol and a bottoms with at least 99.5 mol % water.

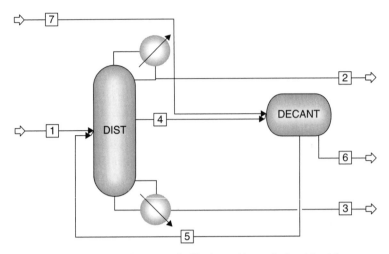

Figure 14.10 Methanol–water distillation with mesityl oxide sidestream.

The feed is available at its bubble point. A sidestream whose purpose is to remove mesity oxide is required. The sidestream will be fed to a decanter with an auxiliary feed of water. Two phases will be formed, one mesity oxide–rich, the other primarily water. The mesity oxide phase will be removed and the aqueous phase returned to the distillation column. A sketch of the process is shown in Figure 14.10. The following is known about the ternary system: No ternary liquid–liquid equilibrium data exist but a heterogeneous azeotrope has been observed; no vapor–liquid equilibrium for the binary mesity oxide–methanol exists. Design the process concepts starting with the steps given in Workshops 14.2a through 14.2h.

Workshop 14.2a A mesity oxide–water azeotrope exits at 760 mmHg and 91.8°C.

MSO in the vapor phase	65.2 wt%
MSO in the liquid phases	93.9 and 2.0 wt%

Determine the van Laar and Uniquac equations $a_{i,j}$ binary parameters.

Workshop 14.2b Estimate the vapor–liquid equilibrium of the methanol–mesity oxide system at one 760 mmHg with a property analysis run using Unifac.

Workshop 14.2c Regress the results of Workshop 14.2b to the $a_{i,j}$ parameters of the van Laar and Uniquac equations.

Workshop 14.2d Fit the vapor–liquid equilibrium data for the methanol–water system of Dunlop (1948) given in Table 14.2 to the van Laar and Uniquac equations using the $a_{i,j}$ parameters.

Workshop 14.2e Develop the liquid–liquid equilibrium tie lines for the ternary system and create a triangular ternary equilibrium diagram at 90°C using the van Laar equation.

TABLE 14.2   Methanol (1)–Water (2) Vapor–Liquid Equilibrium

Temperature (°C)	Pressure (mmHg)	Vapor Mole Fraction, y	Liquid Mole Fraction, x
96.4	760	0.02	0.134
93.5	760	0.04	0.23
91.2	760	0.06	0.304
89.3	760	0.08	0.365
87.7	760	0.1	0.418
84.4	760	0.15	0.517
81.7	760	0.2	0.579
78	760	0.3	0.665
75.3	760	0.4	0.729
73.1	760	0.5	0.779
71.2	760	0.6	0.825
69.3	760	0.7	0.87
67.5	760	0.8	0.915
66	760	0.9	0.958
65	760	0.95	0.975

Workshop 14.2f   Make preliminary runs of the column to aid in locating the feed and sidestream. Experiment with product and sidestream flows to develop appropriate distillation column specifications.

Workshop 14.2g   Link the column and the decanter and develop a solution with loose specifications.

Workshop 14.2h   Refine the specification to develop a final conceptual design.

Workshop Notes

Workshop 14.2a   Aspen Plus provides a unique form of binary parameter equation for almost all activity coefficient equations, which is given in

$$\ln A_{ij} = a_{ij} + \frac{b_{ij}}{T} + c_{ij} \ln T + d_{ij}T + \frac{e_{ij}}{T^2} \tag{14.1}$$

When regressing data, one usually selects $a_{i,j}$ or $b_{i,j}$ or both. The other parameters are rarely used. For a case where only one data point is available $b_{i,j}$ may be used, but extrapolation to other temperatures may cause difficulties.

Workshop 14.2b   A plot of vapor–liquid equilibrium for mesityl oxide–water based on Unifac estimates is shown in Figure 14.11.

Workshop 14.2e   After performing the property analysis, a triangular plot is available from Aspen Plus's Plot Wizard (see Figure 14.12). Note that a methanol–water distillation, without mesity oxide, would take place along the left side of the triangle; however, the presence of a small quantity of mesity oxide places the composition to

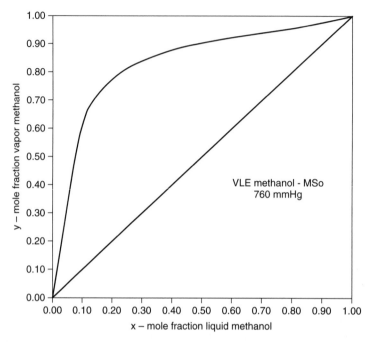

Figure 14.11   Vapor–liquid equilibrium methanol–mesityl oxide at 760 mmHg.

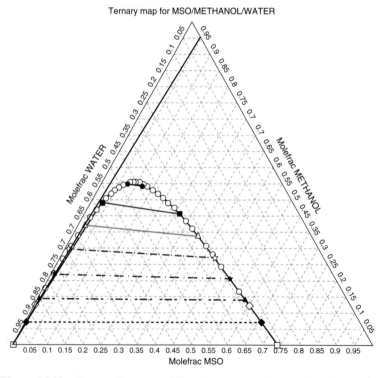

Figure 14.12   Ternary diagram methanol–water–mesityl oxide at 760 mmHg.

the right. The strategy employed here is to locate a vapor composition, within the distillation column, which is to be withdrawn as a sidestream, to the right of a tangent drawn from the left base of the phase envelope along the envelope (see the heavy line on Figure 14.10). When water is added to such a location, the composition of a condensed vapor is within the envelope and forms two phases, one being mesity oxide–rich, which is removed. The aqueous phase is recycled and mixed with the feed to the column.

Workshop 14.2f   Many trial runs are required to estimate the number of stages, the locations of the feed and sidestreams, the value of the reflux ratio, and flow specifications for a product and the sidestream. A good look at the composition profile of the column vapors should provide an estimate of the sidestream location. Use of sensitivity studies will aid in establishing column specifications.

Workshop 14.2g   When linking the column and the decanter, one may experience convergence failure. This may result due to the fact that the flow rate of the recycle stream will vary during the iterations. This is an important factor to take into account when establishing or altering the column specifications to accommodate the recycle stream. For example, a total distillate flow specification may need to be changed to a distillate/feed specification. Occasionally, a block within a process will fail to converge during the iterations but is converged at the conclusion of a run. This is not a problem as long as all blocks and recycles are converged satisfactorily at the conclusion of the run.

Workshop 14.2h   Reducing the reflux ratio and making small adjustments to other column details, such as the number of stages, to the result of Workshop 14.2g will produce a completed conceptual design.

REFERENCES

Arlt, W., Macedo, M. E. A., Rasmussen, P., and Sorensen, J. M., *Liquid–Liquid Equilibrium Data Collection*, Data Series in Chemistry, Vol. V, Parts 1–4, Dechema, Frankfort am Main, Germany, 1979–1987.

Dunlop, J. G., M.S. thesis, Brooklyn Polytechnic Institute, 1948.

Kojima, K., Kato, M., Sunda, H., and Hashimoto, S., *Kagakukogaku*, 32, 3371 (1968).

Seader, J. D., and Henley, E. J., *Separation Process Principles*, 2nd ed., Wiley, Hoboken, NJ, 2006.

Shell Oil Company, Technical Bulletin, Mesityl Oxide.

Stephen, H., and Stephen, T., *Solubilities of Inorganic and Organic Compounds*, Macmillan, New York, 1964.

Van Winkle, M., *Distillation*, McGraw-Hill, New York, 1967.

Wankat, P. C., *Equilibrium Staged Operations*, Prentice Hall, Upper Saddle River, NJ, 1988.

INDEX

Teach Yourself the Basics of Aspen Plus™ By Ralph Schefflan
Copyright © 2011 John Wiley & Sons, Inc.